THE GARNAUT REVIEW 2011

About the author

Ross Garnaut is Vice-Chancellor's Fellow and Professorial Fellow in Economics at the University of Melbourne as well as Distinguished Professor at the Australian National University. He is a Fellow of the Australian Academy of Sciences and an Honorary Professor at the Chinese Academy of Social Sciences. In 2009, Professor Garnaut was awarded the degree of Doctor of Letters, *honoris causa*, from the Australian National University and was made a Distinguished Fellow of the Economic Society of Australia.

Professor Garnaut is currently Chairman of the Papua New Guinea Sustainable Development Program Limited, and its nominee director on the board of Ok Tedi Mining Limited. He is a member of the boards of several international research institutions, including the Lowy Institute for International Policy (Sydney) and AsiaLink (Melbourne). Professor Garnaut was a member of the board of trustees of the International Food Policy Research Institute (Washington DC) from 2003 to 2010 and its chairman from 2006. He was foundation chairman of Lihir Gold Limited from 1995 to 2010.

Professor Garnaut is the author of numerous books, monographs and articles in journals on international economics, public finance and economic development, particularly in relation to East Asia and the Southwest Pacific. Three of his recent books, *The Great Crash of 2008*, *The Garnaut Climate Change Review* and *China's New Place in a World in Crisis*, have been translated into Chinese and published in China by the Social Sciences Academic Press.

Professor Garnaut was the economic adviser to Australian Prime Minister R.J.L. Hawke from 1983 to 1985 and served as Australian Ambassador to China from 1985 to 1988.

Professor Garnaut was appointed as an independent expert adviser to the Australian Parliament's Multi-Party Climate Change Committee in September 2010.

THE GARNAUT REVIEW 2011

Australia in the Global Response to Climate Change

Ross Garnaut

CAMBRIDGE
UNIVERSITY PRESS

CAMBRIDGE
UNIVERSITY PRESS

University Printing House, Cambridge CB2 8BS, United Kingdom

One Liberty Plaza, 20th Floor, New York, NY 10006, USA

477 Williamstown Road, Port Melbourne, VIC 3207, Australia

314-321, 3rd Floor, Plot 3, Splendor Forum, Jasola District Centre, New Delhi - 110025, India

79 Anson Road, #06-04/06, Singapore 079906

Cambridge University Press is part of the University of Cambridge.

It furthers the University's mission by disseminating knowledge in the pursuit of
education, learning and research at the highest international levels of excellence.

www.cambridge.org
Information on this title: www.cambridge.org/9781107691681

The Garnaut Review 2011: Australia in the Global Response to Climate Change
© Commonwealth of Australia 2011

First published 2011

A catalogue record for this publication is available from the British Library

ISBN 978-1-107-69168-1 Paperback

Editorial and artwork by Wilton Hanford Hanover

Index by Michael Harrington

Cover design by Sandra Nobes

Contents

Preface

THE GARNAUT Climate Change Review was commissioned by the Commonwealth, state and territory governments in 2007 to conduct an independent study of the impacts of climate change on the Australian economy. In September 2008, I presented the Review's final report to the Australian Prime Minister. The report examined how Australia, as a single country, was likely to be affected by climate change and how it could best contribute to climate change mitigation, and begin to adapt.

In November 2010, I was commissioned by the Australian Government to provide an update to the 2008 Review. In particular, I was asked to examine whether significant changes had occurred that would affect the key findings and recommendations reached in 2008.

The commissioning of the update reflected the changed international and domestic landscapes for climate change action following the international climate change conferences in Copenhagen and Cancun in 2009 and 2010, and the Great Crash of 2008. What implications did these events have for climate change policy globally and in Australia?

This book, the final report of the update process, is the product of seven months of careful research, analysis, expert studies and consultation, which have examined key developments in the past two and a half years across a range of areas—the climate science, global greenhouse gas emissions, international progress on climate change mitigation, Australia's land and electricity sectors, innovation and technology, and carbon pricing. Eight detailed update papers were released between February and March 2011. Two supplementary notes came out at the same time as this book. These materials and other supporting information can be found on the Garnaut Climate Change Review website at www.garnautreview.org.au.

Ross Garnaut

Melbourne

31 May 2011

Introduction

I WAS EXPLAINING to the Multi-Party Climate Change Committee early in 2011 how I had worked out the costs and benefits of reducing emissions for the 2008 Review. The costs of reducing emissions will come straightaway. The benefits of reducing damage from climate change will come later—many of them to later generations of Australians. In fact there will be more and more benefits for later and later generations. So I needed a way of comparing the value of income to Australians who are alive right now with incomes of young Australians later in their lives and Australians who are not yet born.

'So we had to choose the right discount rate', I said. 'We can't use the discount rates that determine values in the share market, because they take into account risks of a kind that are not relevant here.'

I got the feeling that the mention of discount rates had set Prime Minister Gillard's mind towards what she would say to Hillary Clinton about Afghanistan, Bob Brown's to the grandeur of the Styx Valley, and Tony Windsor's to the good rain that was falling on the Northern Tablelands.

But then I said something that brought back the prime minister's attention.

'If we used the share market's discount rate to value the lives of future Australians', I said, 'and if we knew that doing something would give lots of benefits now but would cause the extinction of our species in half a century, the calculations would tell us to do it.'

The beginnings of a smile on her face became a hearty laugh.

'You've got us there, Ross', she said, as the others were infected by the lift in spirits and joined the laughter. 'That's a unanimous decision of the committee. We're all against the extinction of the human species.'

The 2008 Garnaut Climate Change Review compared the costs and benefits of Australia taking action to reduce the damage of climate change caused by humans. It concluded that it was in Australia's national interest to do its fair share in a strong global effort to mitigate climate change.

The 2008 Review accepted the central judgments from the mainstream science about the effects of changes in greenhouse gas concentrations in the atmosphere on temperature, and about the effects of temperature changes on climate and the physical earth. I formed the view that the mainstream science was right 'on a balance of probabilities', and that errors were as likely to be in the direction of understatement of damage to human society as in the direction of overstatement.

I used the results of the science to model the impacts of climate change on the Australian economy, including impacts on agricultural productivity, our terms of trade, and infrastructure. The model included links to the global economy and was based on Australia doing its fair share in a global effort to reduce the damage from climate change.

The modelling showed that the growth rate for Australian national income in the second half of the 21st century would be higher with mitigation than without. The present value of the market benefits this century fell just short of the value of the costs of mitigation policy. However, when we took account of the value of Australians' lives beyond the 21st century, the value of our natural and social heritage, health and other things that weren't measured in the economic modelling, and the value of insuring against calamitous change, strong mitigation was clearly in the national interest.

New developments

And so we come to today. The purpose of this book is to examine how developments in science, diplomacy, political culture and the economy have affected the national interest case for Australian climate change action.

Since the 2008 Review, the science of climate change has been subjected to intense scrutiny and has come through with its credibility intact. The findings continue to be sobering. Unfortunately, new data and analysis generally are confirming the likelihood that outcomes will be near the midpoints or closer to the bad end of what had earlier been identified as the range of possibilities for human-induced climate change.

Global average temperatures have continued to track a warming trend. The year 2010 ranked with 2005 and 1998 as the warmest on record, with global average temperatures 0.53°C above the 1961–90 mean. For Australia, 2009 was the second-warmest year on record and the decade ending in 2010 has easily been Australia's warmest since record keeping began.

I noted in the 2008 Review the curious Australian tendency for dissenters from the mainstream science to assert that there is no upward trend in temperatures, or that if there had been a warming trend it has ceased or moved into reverse. Such assertions were prominent in some newspapers and blogs, but also appeared in serious policy discussions. The assertions were curious because the question of whether the earth is warming or not is amenable to statistical analysis.

It so happens that answering questions of this kind comes with the professional kitbag of economists who work on statistical analysis of series of data that cover periods of time. For the 2008 Review, I asked two leading Australian econometricians who are specialists in this area, Trevor Breusch and Fashid Vahid, to analyse the data on temperature. Their conclusion was clear. There is a statistically significant warming trend, and it did not end in 1998 or in any other year. I had the analysis repeated with three more years of data for this book, with the same conclusions.

New observations of a changing climate include an increase in extreme weather events. The Black Saturday fires in Victoria in 2009 and recent major cyclones in Queensland are both consistent with expected outcomes in a warming world, although we cannot draw conclusions about direct cause and effect.

Other studies since 2008 have confirmed that Australia is also seeing historically unprecedented periods of wet and of dry in different areas of the continent.

Globally, rising sea levels continue to track the upper levels of modelling. Considerable debate is under way about the causes and potential extent of sea-level rise. The latest research suggests that, beyond the effects of thermal expansion, the melting of the great icesheets of Greenland and West Antarctica may contribute much more than was previously thought to sea-level rise. The debate is unresolved but oriented towards higher not lower outcomes.

New research has also contributed to our understanding of 'tipping points' in the climate system. These are points at which warming of the climate triggers irreversible damage and a feedback loop for further warming. The new research has focused on identifying and testing potential early warning indicators of an approaching tipping point.

Progress has also been made on ruling out other possible causes of warming, such as changes in the amount of solar radiation reaching the earth. Scientists have identified 'fingerprints' of warming that confirm human influence. A primary example is the pattern of warming in the layers of the atmosphere. Under increased greenhouse gas scenarios, climate models predict that the lowest layer of the atmosphere (the troposphere) should warm, while the next layer up (the stratosphere) should cool. This has been confirmed by recent observation. If increased output from the sun were the cause, both layers could be expected to warm. These developments and more are examined in Chapter 1.

Since 2008, advances in climate change science have therefore broadly confirmed that the earth is warming, that human activity is the cause of it and that the changes in the physical world are likely, if anything, to be more harmful than the earlier science had suggested. This has led me to shift my judgment about the reputable science from being right 'on a balance of probabilities' to 'beyond reasonable doubt'.

Chapter 2 focuses on likely amounts of greenhouse gas emissions in the absence of mitigation. It examines the effect on emissions of the big global economic developments following the global financial crisis—the Great Crash of 2008.

Emissions under business as usual are on a somewhat lower trajectory in the developed countries, mainly as a result of the loss of growth momentum after the Great Crash. This is roughly balanced in the period to 2030 by continued strong growth in the developing countries.

The result is a global emissions trajectory in the event of business as usual that is little changed from 2008, but is constitutionally very different. The share of emissions growth attributed to large developing nations like China and India has grown as developed countries' growth has shrunk.

Australia is an exception among the developed countries. Following the Great Crash, Australia's rich endowment of natural resources has helped fuel the outstanding growth in the large developing countries. The resulting high terms of trade project a strong growth performance based around high levels of investment in mines, including for coal and gas. The projection of Australia's emissions trajectory without mitigation to 2020 has grown to 24 per cent above 2000 levels—4 per cent above the levels expected in 2007—despite new policy measures in the intervening years.

The shift of the centre of gravity of growth towards developing countries is wonderful for human wellbeing so long as we can manage the consequence: that mitigation becomes more difficult. By 2030, the average income in developing economies will be slightly more than a quarter of that of the United States. The potential for further catch-up growth in incomes and emissions is stark.

However, there has been a major positive development. The world has already moved considerably beyond the business-as-usual case described above. Chapter 3 examines important developments in the global framework for action that give hope of holding global emissions to levels that avoid dangerous climate change.

The 2009 Copenhagen and 2010 Cancun conferences of the United Nations Framework Convention on Climate Change led to an important new direction in global mitigation policy. The diplomatic fiasco of the

Copenhagen conference disguised a breakthrough new agreement that addressed the great failing of the Kyoto Protocol. It incorporated mitigation targets for the United States and the large developing economies, notably China. All countries also agreed to contain global warming within 2°C.

The Copenhagen agreement had its weaknesses. The new targets were voluntary, not ruled by legal obligation and delayed the prospect of the trading of carbon permits between countries. But they did establish a new 'pledge and review' system that included new mechanisms for measuring and tracking emissions.

The meeting at Cancun consolidated and extended the new agreement, as well as the mitigations targets pledged by developed and developing countries.

The pledged targets of all countries that play substantial roles in global emissions are evaluated in Chapter 4. The ranges for the United States, the European Union and Japan together correspond to entitlements for the early stages of a moderately ambitious—if not strong—global agreement. On average, developed countries' pledged 2020 targets are somewhat less ambitious than are needed to hold the concentration of greenhouse gases in the atmosphere to 550 parts per million (ppm) of carbon dioxide equivalent.

For developing countries, targets are measured not in absolute reductions but in reductions in emissions intensity. The modified contraction and convergence framework described in the 2008 Review implied a targeted reduction in China's emissions intensity of 35 per cent between 2005 and 2020 if global concentrations of carbon dioxide were to be limited to 450 ppm. At Copenhagen and Cancun, China pledged to reduce its carbon intensity by 40 to 45 per cent between 2005 and 2020.

China has already achieved considerable success in the implementation of its pledged targets with sweeping regulatory actions in energy and innovation. Chinese leaders have been pleasantly surprised at the pace and cost of change and are growing in confidence that they will later be in a position to offer more aggressive pledges still.

In this new world of concerted unilateral action, countries closely examine each other's efforts to confirm that each is contributing its fair share. Freeloading may contribute in only a small way to overshooting global targets, but it threatens the entire global effort as all countries look to one another for reassurance that the pledged progress is being made.

Solutions

So, developments in science, global emissions profiles and shifts in the structure of global climate change agreements have all strengthened the national interest case for a stronger Australian mitigation effort.

What domestic policy response should we take? Once we know what our fair share is in the global effort to reduce greenhouse gas emissions, we can work out how to do it at lowest cost. This exercise was undertaken in detail and with great care for the 2008 Review. There are two basic approaches to achieving the required emissions reduction: a market-based approach, built around putting a price on carbon emissions; and a regulatory approach, or direct action.

In the market-based approach, carbon can be priced in two ways. Fixed-price schemes, or carbon taxes, set the price and the market decides how much it will reduce the quantity of emissions. Floating price schemes set the quantity of emissions and permits to emit are issued up to that amount. The permits are tradeable between businesses and so the market sets the price. There are various hybrid approaches that combine fixed prices for a period with floating later on, and floating prices at some price levels with a price floor or a price ceiling or both.

In the alternative route, regulation or direct action, there are many ways that government can intervene to direct firms and households to go about their business and their lives. The Chinese Government's direct action includes issuing instructions for factories with high emissions to close, subsidising consumers who buy low-emissions products like solar electricity panels and electric cars, and restricting new investment in industries judged to have undesirably high emissions.

Chapter 5 explores these options and argues for a three-year fixed carbon price followed by a carbon trading scheme with a floating price. This confirms the approach proposed in the 2008 Review for circumstances similar to those in which we now find ourselves. This is Australia's best path forward towards full and effective participation in humanity's efforts to reduce the dangers of climate change without damaging Australian prosperity.

One distinct advantage of addressing climate change mitigation through a market-based carbon price is that it raises considerable revenues. These can be used to buffer the transition to a low-carbon economy for Australian households on low and middle incomes, as well as to offer security to the most vulnerable low-income households.

A carbon price of $26 will raise approximately $11.5 billion in the first year and rise over time. Efficiency and equity objectives would be best served by allocating the majority of this revenue to households, perhaps modelled on the kind of tax and social security reforms envisioned in the Henry review.

At the same time, slices of this revenue should also be used to support innovation in low-emissions industries, provide incentives for biosequestration in rural Australia and prevent export industries from being placed at a disadvantage against international competitors that are not yet subject to comparable carbon constraints. Chapter 6 is a national interest analysis of how compensation should be deployed to each of these groups.

Of course, under a direct action or regulatory approach, costs are imposed on households and businesses but none of these benefits are available to balance them.

National versus vested interests

Yet, as clear as the case for carbon pricing may seem, the political basis for such policies has weakened since 2008. Alongside the central discussion of climate policy, this book is a guide to another struggle that is deeply colouring the climate change debate—the struggle between special interests and the national interest.

This conflict is not new. Indeed, it is always with us, and always will be. But there are periods when the special interests have had the strongest hold on policy, and others in which policy making is strongly grounded in the national interest.

It is salutary to recall that Australia, with New Zealand, had the poorest productivity performance of all the countries that are now developed through the 20th century to the mid-1980s. The long period of underperformance had its origins in the domination of policy by business and union vested interests. Political leaders responded to democratic pressures with protection and regulation. There was little competition to prompt firms to seek new, more productive ways of doing business.

We managed to break out of that from 1983 onwards, and entered a remarkable period of productivity-raising reform. After a while, suggestions for policy reform were not taken seriously by anyone unless they were placed in a sound national interest context. The leadership of the Australian Council of Trade Unions responded quickly to the circumstances offered by a new approach to government. To remain relevant to the policy process, the old, protectionist business lobbies were reformed as the Business Council of Australia.

Protective and regulatory constraints on higher productivity were progressively reduced.

The period of policy reform oriented to the national interest lasted until the turn of the century. Productivity responded to the new political culture and the policies that it supported. Australian productivity growth in the 1990s after the recession at the start of the decade was the highest in the developed world.

The end of the era of reform can be dated fairly precisely. No major market-based productivity-raising reform has survived the political process since the tax reform package of 2001. That package was itself deeply compromised by the increased distortions in federal–state financial relations that had been introduced as the political price for reform. And it was bought with 'overcompensation' amounting to about a percentage point of Australian national income.

From the beginning of the 21st century, Australian policy making has reverted to type. Business and union organisations refocused on securing sectional gains. Governments responded. There could be no policy change if there were any losers, so there could be no productivity-raising change at all. There has been little increase in the productivity with which resources (capital and labour together) are used in Australia so far in the 21st century, and none at all since 2003.

The absence of total productivity growth over the last decade was covered up for a few years at the beginning of the century by an extraordinary boom in housing and consumption, mainly funded by unsustainable foreign borrowing by our banks. That boom would have ended quickly in tears had we not been rescued by a resources boom—much higher export prices and, after a while, investment in resources—of historic dimensions. Now it will end in tears after a longer period.

This is the problematic political context of the climate change policy discussion.

Some business leaders have recently drawn attention to the need for long views and hard decisions in policy making. They say that the minority Labor government elected by the Australian people in 2010 is weak and lacks long time horizons.

A more accurate accounting would recognise that the current government has taken on the most difficult and long-dated policy reform that has ever been attempted. It has taken on a reform in the national interest that must overcome stronger pressures from sectional interests than any since the contests over protection in the 1980s and early 1990s. That part of big business that is active in the debate has taken on the role of spoiler. Chapter 7 examines this phenomenon and notes that in a political economy already dominated by vested interests, a transparent, market-based carbon price is far less likely to be unduly influenced by private interests than a regulatory approach which provides recurring opportunities for lobbying. A market-based approach will, for this among other reasons, cost Australians substantially less.

The same calculation applies to adapting to the degree of climate change that is already locked in regardless of mitigation efforts from this time forth. Chapter 8 looks at the likely adaptation measures that will be required. The key to success and greatest efficiency will be maintaining a productive, flexible, market-oriented economy.

The independent centre

I noted in the 2008 Review that the diabolical policy issue of climate change had a 'saving grace' that may make all the difference—that climate change is an issue in which a high proportion of Australians are deeply interested. This provided an opportunity for the exercise of authority by an independent centre, against the claims of interests that see themselves as being negatively affected by mitigation. My consultations and community engagement through the update of the Review have confirmed the continued presence of the saving grace, although it has been tested by the bizarre quality of the public discussion of recent times.

In confronting the spoiling voices, we must remember that rejection of current proposals for carbon pricing would not end the debate over climate change policy. It might, however, end the possibility of action at relatively low cost.

The increasing impact of climate change as well as policy developments abroad would prompt continued pressure for new policy in Australia. Inaction by Australia, with the highest emissions per person in the developed world, would invite retaliation in trade and other areas of international cooperation. If current efforts on carbon pricing failed, debate would continue over how much Australia should do and how we should do it. This would continue to raise the supply price of investment in businesses that might be affected by restrictions on emissions. The political system would respond to continued community interest in and pressure for action on climate change by myriad costly interventions. The failure of current efforts to place a price on carbon through much of the economy would open the way to a long period of policy incoherence and instability.

There is no reason why carbon pricing should continue to be a matter of partisan political division in Australia. In much of the world—perhaps everywhere except Australia and the United States—concern for global warming is a conservative as much as a social democratic issue. The conservative governments of Germany, the United Kingdom, France and the Republic of Korea are playing important global leadership roles. Even in the United States, the most effective political leadership on climate change has come from a Republican governor of California and a Republican mayor of New York.

A concern to avoid dangerous climate change fits naturally within the conservative tradition. It may be rational for the radical to risk the institutions of human civilisation in a throw of the climate change dice, just as Lenin saw merit in inflation in the capitalist countries. The radical may hope that the outcome will open the social and political order to new shapes. It is strange for the conservative to embrace such risk.

Nor do the characteristic divisions between the conservative and social democrat argue for conservative opposition to carbon pricing. Market-based approaches to mitigation sit as easily with a conservative party that is self-described as liberal, as they do with social democratic parties.

It would be open to current or future leaders of the conservative side of Australian politics to take over ownership of carbon pricing arrangements once they are in place. The interests of their future governments, as well as those of Australia, would be served well by the continuation of carbon pricing.

Transformations

The Member for New England in the House of Representatives, Tony Windsor, has commented that if the whole world really were doing nothing, there would be no point in Australia seeking to reduce greenhouse gas emissions. We might as well join the other lemmings as they rush over the high bluff.

Fortunately for humanity—and in particular for Australians as residents of the country in the developed world that is most vulnerable to climate change—much of the rest of the world is not behaving like lemmings.

Despite the raucous disputation and associated inaction in Australia, other countries have kept alive the possibility of effective global action. There is substantial action in many countries to constrain greenhouse gas emissions, but the future shape of international action could evolve in a number of different ways. Australian policy should seek to shape that evolution in line with our national interest in effective mitigation of climate change, while calibrating Australian policy to what others are doing.

Both the Australian Government and the Opposition have committed themselves to a minimum reduction of emissions of 5 per cent by 2020. This book defines a process through which we would adjust that share over time in light of what others were doing.

If we commit ourselves to doing our fair share, and maintain that level of commitment through the governance mechanisms recommended in this book, there can be a smooth adjustment to increased international effort. The targets would be tightened as other countries became more ambitious in reducing emissions. Carbon prices would rise on international markets and that would be reflected in the Australian price. There would be certainty for business about the process, although the carbon price would change over time. But price fluctuations are the kind of uncertainty with which business is familiar—like the uncertainties in commodity and financial markets that are managed in the normal course of business.

How much the transition costs depends on Australians' success in innovation. The carbon price will make it profitable to do new things in new ways. Some Australian businesses and individuals will do those things and fund those ways, and others will learn from them. We need a lot of technological change over a short period of time. Chapter 9 discusses policies to make sure we get it.

The effect of the carbon price upon the two industry sectors that are most enmeshed by climate change and mitigation—agriculture and electricity—are covered in chapters 10 and 11.

The Australian rural sector will be challenged greatly by climate change, which will generate higher prices for farm products but place barriers against making good use of them. A world of effective global mitigation would provide many opportunities for Australian farmers, as they would be in a better position to take advantage of higher world prices resulting from other developments in the global economy. Farmers should be able to sell the full range of legitimate biosequestration credits into the carbon pricing scheme, providing the basis for a new industry of considerable potential.

The evolution of the electricity sector under carbon pricing should not cause the community anxiety. Australia has an incomparable range of emissions-reducing options. The early stages of the transition will see expansion of gas at the expense of coal alongside the emergence of a range of renewable energy sources. The carbon price will arbitrate between the claims of different means of reducing emissions as the profitability of each is affected by many domestic and international developments. Whether or not coal has a future at home and as an export industry depends on the success of technologies for sequestration of carbon dioxide wastes. There is little reason for concern about the physical security of energy supply during the transition to a low-emissions economy, but I propose some cost-effective measures to ease anxieties in parts of the community.

This book is the story of Australia's national interest in contributing our fair share to a global mitigation effort. It is a story of how market-based approaches to mitigation can bring out the best in Australians, and a return to regulatory approaches the worst. Both best and worst lead us to the same conclusion: that a broad-based market approach will best preserve Australian prosperity as we make the transition to a low-carbon future.

PART I
THE GLOBAL SHIFT

1 Beyond reasonable doubt

THE INTERNET is a wonderful research tool. With the click of a mouse, you can beam yourself into what seems an infinite number of important lectures by eminent thinkers around the world.

Just such an opportunity awaits if you search for Richard A. Muller, professor of physics at the University of California, Berkeley. The results of such a search include a video of Professor Muller castigating the scientists at the centre of the 2009 Climategate scandal.

'Climategate' is the name given by sections of the American media to the 2009 imbroglio surrounding leaked emails from the Hadley Centre in the United Kingdom. The emails were used to suggest that some scientists had been selective in their use of data to support the idea of global warming. In the video, Professor Muller berates the Hadley Centre scientists for smoothing data to produce alarming graphs that would make global warming 'incontrovertible' to the public. Professor Muller concludes by announcing his own major study into the measurement of global warming, the Berkeley Earth Project, without 'the bias', he says in the video.

Some months later, in March 2011, Professor Muller appeared before a US congressional committee at the invitation of Republican members opposing action on climate change. He was there to present the preliminary results of his bias-free project. To his surprise, he said, and certainly to the surprise of his hosts, the results of the project tallied very closely with those of the Hadley Centre's temperature measurements, as well as those of the US National Aeronautics and Space Administration and the National Oceanic and Atmospheric Administration.

The fact is, that despite human imperfection, modern science on climate change has held up well under withering scrutiny. The vast majority of those who have spent their professional lives seeking to understand climate and the impacts of human activity on it have no doubt that average temperatures on earth are rising and that human-induced increases in greenhouse gases are making major contributions to these rises. They are supported in this by the learned academies of science in all of the countries of scientific accomplishment.

Where dissent is found in the community of scientists with genuine climate credentials, it is among a small number who argue that the effects of increases in greenhouse gases are small compared with other sources of changes in temperature.

But a larger number of alternative views can also be found on the other side of the debate. There are reputable scientists who argue that great changes in climate are triggered by lower greenhouse gas concentrations than the mainstream science suggests.

There are other important debates in the scientific community about the impacts of rising temperatures. For example, scientific climate models reveal wide variations in the regional distribution of projected changes in rainfall.

Another example is the extent of sea-level rise that is likely to be associated with specified degrees of warming. The decisive research relates to the mass of land-based ice in Greenland and Antarctica. This is a large issue, as the complete melting of Greenland ice would raise sea levels by about 7 metres, of west Antarctica by about 6 metres, and of east Antarctica by much larger amounts. The mainstream view from the peer-reviewed literature, brought into the public domain through the 2007 Intergovernmental Panel on Climate Change (IPCC) Fourth Assessment Report, argued that sea-level rise would result from expansion of the oceans' volume as ocean temperatures rose and from the melting of alpine glaciers. It also included contributions from the surface melting of land-based ice in Greenland and Antarctica but not the potential losses from dynamical processes—the calving of large icebergs from outlet glaciers. This was not because the scientists with relevant expertise didn't think that these processes were important, but because not enough was known about them to include them in models of sea-level rise.

During the early research for this book, it was disconcerting to find that the few deep specialists in land-based ice expressed the view privately that there *would* be a major contribution from dynamical processes in Greenland and west Antarctica to sea-level rise this century. The dimensions of the contribution are uncertain, but they are certainly substantial and possibly greatly disruptive. All declined to put their private views on the public record.

The end point of the four-year research process that produced this book is the conclusion that it is highly probable that the central proposition of the mainstream science is correct. Most of the global warming since the mid-20th century is very likely due to human-caused increases in greenhouse gas concentrations. Furthermore, the range of genuine scientific views from the peer-reviewed mainstream suggests that temperatures and damage from a specified level of emissions over time will be larger than is suggested by the middle ground of the mainstream science.

The carbon cycle

Carbon is transferred, in various forms, through the atmosphere, oceans, plants, animals, soils and sediments as part of the carbon cycle. The term 'carbon budget' is often used to describe the balance of inflows and outflows that lead to the accumulation of carbon dioxide in the earth's atmosphere. These natural inflows and outflows were approximately equal for several thousands of years before the effects of the industrial revolution became apparent around 1800.

Since the early 19th century there has been a large and increasing inflow of carbon dioxide into the atmosphere from human activities. The burning of fossil fuels, cement production and other industrial processes, as well as deforestation or land clearing, are largely the cause.

Emissions from fossil fuels are the largest source of atmospheric carbon dioxide from human activities. Carbon dioxide emissions from fossil fuel combustion increased by about 2 per cent per year in the 1970s and 1980s, and by only around 1 per cent in the 1990s. Between 2000 and 2008, the annual increase in fossil fuel emissions grew to 3.4 per cent.

This trajectory is well above the IPCC scenario with the highest emissions through to 2100, which had been considered to be extreme until the publication in the 2008 Review of more realistic assessments. It is tracking closely the projections presented under business as usual in the 2008 Review. Even with a recent slight drop in the annual rate of increase due to the Great Crash, the average increase in emissions for the last decade was around 3 per cent.

Land-use changes, such as deforestation and conversion to crops, are the second-largest source of carbon dioxide emissions from human activities. In contrast to the 29 per cent increase in fossil fuel emissions between 2000 and 2008, land-use change emissions have been fairly steady and now account for less than 15 per cent of total emissions.

The human-caused increase in carbon dioxide in the atmosphere is partly offset by natural carbon dioxide 'sinks' in both the land and oceans. The efficacy of land-based carbon sinks is determined by the balance between plant growth, respiration from plants and soils, and land-use disturbances, such as fire and forest clearing. The ocean acts as a carbon sink because carbon dioxide dissolves in ocean waters when concentrations in the atmosphere are higher than those at the ocean's surface. This dissolved carbon is moved into the deeper ocean by overturning currents, and also by the sinking of dead organisms.

Over the past 50 years, the uptake by these natural sinks has continued to remove around half of the carbon dioxide put into the atmosphere, despite the increasing human-caused emissions. The carbon is taken up in roughly equal proportions by the land and the oceans. There is considerable variation in the strength of these natural sinks from year to year, largely in response to climate variability.

Some recent studies have indicated that there has been a decline over the last five decades in the percentage of carbon dioxide emissions from human activities that is absorbed by natural carbon sinks. There have been suggestions that this shows that natural carbon sinks are slowly 'losing the race' against the rapidly growing human-caused emissions. But there is controversy in the scientific community over these results.

The magnitude and the rate of the increase in concentrations of carbon dioxide, methane and nitrous oxide in the atmosphere in the last century have increased considerably compared to the past millennium. Between the years 1000 and 1750, carbon dioxide concentrations ranged between 275 and 285 parts per million (ppm). It then took more than 200 years, until the 1970s, for concentrations to increase by 50 ppm, but only another 30 years for a further increase of about 50 ppm to the current levels. Carbon dioxide concentrations have increased from 379 ppm in 2005 to 390 ppm in early 2011.

Concentrations of the two other main greenhouse gases—methane and nitrous oxide—have also increased, and remain well above concentrations of the last 20,000 years. Methane concentrations have more than doubled since the industrial revolution and increased by about 30 per cent in the 25 years up to the IPCC Fourth Assessment Report, although with some recent reductions in the rate of growth. The increase in methane concentrations is probably due to increased methane emissions from high latitudes and tropical wetlands, linked to increases in global temperatures and tropical precipitation. The concentration of nitrous oxide is now about 18 per cent above the 1750 level.

Between 1998 and 2010, the increase in greenhouse gas concentrations in the atmosphere is equivalent to a change in carbon dioxide concentrations from 438 ppm to 465 ppm.

Temperature trends

One of the IPCC's main conclusions in its 2007 report was that the 'warming of the climate system is unequivocal'. Global average temperatures had risen considerably since measurements began in the mid-1800s, and since pre-industrial times (1850–99) the global surface temperature had increased by

0.76 +/- 0.19°C. The Royal Society recognises that there is wide agreement in the scientific community on this aspect of climate change.

The World Meteorological Organization concluded: 'The year 2010 ranked as the warmest year on record, together with 2005 and 1998. Data received by the WMO show no statistically significant difference between global temperatures in 2010, 2005 and 1998. In 2010, the global average temperature was 0.53°C … above the 1961–90 mean.'

The IPCC's 2007 conclusion about warming trends was not based only on surface temperature data, but also on the changes in other levels in the atmosphere. Trends in other areas of the climate system, such as the uptake of heat by the oceans and the melting of land ice, such as glaciers, are also occurring. Hence, there is wide-ranging evidence of a warming trend in different indicators produced by independent researchers that provides a consistent story of a warming world.

In Australia, annual average temperatures have increased by 0.9°C since 1910. Figure 1.1 shows Australian average temperature anomalies since 1910. While 2005 is still the hottest year on record based on the mean annual temperature across Australia, 2009 was the second-warmest year.

Figure 1.1: Australian annual average temperature anomalies, 1910–2010

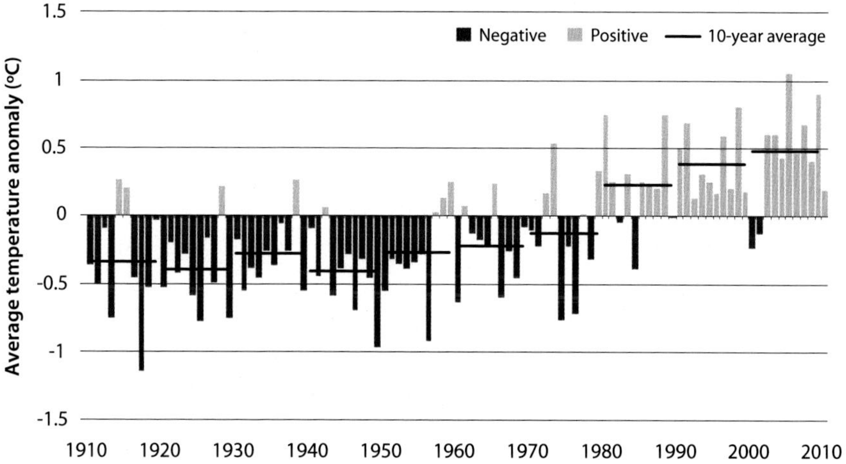

Note: The data show temperature difference from the 1961–90 average.

Source: Bureau of Meteorology time series data, retrieved 10 February 2011.

The decade ending in 2010 has easily been Australia's warmest since record keeping began. It continues a trend of each decade being warmer than the previous that extends back to the 1940s. The milder year in 2010 demonstrates that individual years can still be relatively cool even as the warming of Australia's climate continues.

New climate observations

When reporting on newly observed changes in the climate that have accompanied increases in temperature, we must remember that we are looking at a relatively short period (the research for this book was conducted over four years from 2007). There is inevitably a focus on recent weather and extreme events. Any set of observations over a short period will reflect the dynamic nature of the climate. Apparently random fluctuations from the norm create 'noise' that can make longer-term patterns and trends difficult to identify over a short period. Rather than being viewed as indicative of a change in climate or otherwise, single events or annual data must be considered within the context of the growing dataset of climate information.

So, in the future, it is quite consistent with the strictures of scientific observation of climate change to expect the climate system to respond in variable ways to an increased concentration of greenhouse gases.

Years like 2010 will continue to occur, where temperatures were high globally but some countries (in this case Australia) were relatively cool.

The regional variability of climate change will also manifest in severe weather events of an intensity that is rare at a particular place and time of year. 'Severe weather events' include (among others) heatwaves, heavy rainfall and floods, droughts, tropical cyclones and bushfires.

While it is difficult to attribute specific causes to individual severe weather events, climate change is expected to increase the risk of extreme events. The changes include greater frequency (heatwaves, bushfire conditions, floods, droughts), greater intensity (all of these plus cyclones) and changes in distribution (average rainfall).

The potential impact of climate change on severe weather events has been brought to the fore recently due to a series of major climate events globally and in Australia. Individual events may be assessed for their consistency with expectations in a warmer world and compared with the equivalent expectations if the underlying climate conditions had not been changing. For example, the conditions of the 2009 Black Saturday fires in Victoria were consistent with expectations for a warming world. There will be an increase in the frequency of such conditions as the world continues to warm.

However, such comparisons generally do not allow us to state categorically that such an event could only have occurred with climate change. We can say that the extreme conditions that were the backdrop to the Victorian bushfires or the 2011 Queensland cyclones and floods will be more likely to occur and will occur more often in a warmer world.

In making assessments about the effects of warming on extreme events, we should keep in mind that we are only in the early stages of global warming. Land temperatures have increased by less than half of the level that would be expected even with effective, strong mitigation to hold greenhouse gas concentrations to 450 ppm carbon dioxide equivalent. They have increased only one-quarter of the rise expected in the event of partially successful mitigation to 650 ppm carbon dioxide equivalent. And the increase to date is only a small fraction of ultimate temperature rises in the event of no mitigation at all.

The strength of some severe weather events is likely to rise more than proportionately with the increase in average global temperature. It is therefore to be expected that the reflection of global warming in severe weather events is in an early and weak stage.

Some recent work looking at events in the northern hemisphere has advanced understanding of the probability of a link between extreme events and climate change. A recent study looked at the probability of human-induced climate change increasing the risk of an extreme autumn flood event that occurred in the United Kingdom in 2000.

To analyse this single flood event, thousands of simulations of the weather experienced at the time were generated under realistic conditions, and also under conditions where the warming influence from greenhouse gas emissions had been removed. In nine out of ten cases, the results showed that human-induced greenhouse gas emissions in the 20th century increased the risk of the flood event by more than 20 per cent, and in two out of three cases by more than 90 per cent.

Another study used a similar approach to severe rainfall and heavy snowfall in the northern hemisphere. It found these events could not be explained without factoring in the increases in greenhouse gases from human activity.

A recent study on Australian temperature and rainfall records between 1911 and 2008 investigated changes in the percentage area of the continent experiencing extreme cold, hot, dry or wet conditions. It showed that for Australia as a whole—not at all locations—there has been an increase in the extent of wet extremes and a decrease in the extent of dry extremes, both annually and during all seasons.

Historically, co-variations in Australian extremes have been either hot and dry, or cold and wet. The same study detected a long-term shift towards wet extremes and hot extremes occurring at the same time, which is not consistent with processes causing variability between years and decades.

This suggests that the long-term trends are influenced by a separate process. The increase in both hot and wet extremes is consistent with changes as a result of increased concentrations of greenhouse gases.

One of the more obvious severe weather events on the wet side of the climate ledger is tropical cyclones. With more heat energy in the atmosphere and oceans, there will be fewer cyclones overall but more cyclones with extreme force. Tropical cyclones occur when warm, moist air rises and then condenses, leading to the release of energy and the formation of wind. Tropical cyclones do not form unless the sea surface temperature is above 26.5°C. Theory and modelling suggest that, as oceans warm, there will be more energy for conversion into tropical cyclone wind, leading to increased wind speeds and more intense cyclones.

Analysis has shown that rainfall associated with tropical cyclones (within 300 kilometres) is likely to increase by 17 per cent on average by 2070 compared to 1980. The same study showed that a larger percentage of tropical cyclones will produce higher wind speeds in 2070 than in either 1980 or 2030. These regional findings are consistent with recently published international studies.

The El Niño – Southern Oscillation will have a significant effect on future cyclones and storms in Australia, so it is difficult to project changes in the frequency and intensity of cyclones without a better understanding of this phenomenon.

Conversely, a considerable body of Australian research suggests that the persistent dry conditions in parts of the south-west and south-east of Australia are at least in part due to climate change. In the south-west, the movement of autumn and winter rain-bearing weather fronts to the south is associated with a southward shift of a large-scale atmospheric circulation system that has been linked to climate change. In the south-east, decreasing rainfall is strongly associated with strengthening over the region of high surface pressure that causes much of the seasonal variation in weather in the south of the continent. This increasing pressure is consistent with the rise in global mean temperature, and expectations from the physics of the climate system.

Climate models indicate that as temperatures rise further, rainfall will increase close to the poles and in equatorial regions, and decrease in subtropical and some temperate regions. Climate change will also influence the seasonal and daily patterns of rainfall intensity. The risk of drought is expected to increase in the mid-latitudes (southern Australia). Increased flood risk is also expected as rainfall is concentrated into more intense events.

Unlike future temperature, which is always simulated to increase throughout Australia, the results from some climate models show that many locations could be drier, while others suggest those locations could be wetter. However, the majority of climate models project a drier future for southern Australia than was experienced last century.

The 2008 Review noted research showing that up to 50 per cent of the decline in south-west Western Australia's rainfall was due to human-induced climate change. A reduction in rainfall results in a proportionately larger fall in stream flows. Annual inflows to Perth's water storages have declined markedly since the 1970s, and since 2006 have been only 17 per cent of the long-term average before this observed decline. Inflows reached a record low in 2010, which was the driest year on record for south-west Western Australia. Climate change is likely to contribute to further reductions in surface water availability in southern Australia.

Changes to oceans and sea levels

The world's oceans store the majority of heat within the climate system. As a result, changes in the heat content of the oceans are a critical element in climate change, leading to increased sea surface temperatures and contributing to changes in sea level. Analysis of historical observations confirms that the oceans have warmed since 1950 and that they have stored more than 90 per cent of the increase in heat associated with global warming. This warming has continued over the last 15 years.

As any high school student should be able to tell you, as water warms it also expands in volume. When applied to the oceans, this process—called thermal expansion—results in rising sea levels. The total observed sea-level rise during the 20th century was about 160 millimetres, averaged over the global oceans. More recent observations indicate that sea level has been rising more rapidly over the past two decades, with the average rates since 1993 about 3.2 millimetres a year.

The sea-level rises from 1990 were anticipated in the two most recent IPCC reports. Observed sea level is tracking near the upper limit of the IPCC's 2007 projections for sea-level rise, as shown in Figure 1.2.

Figure 1.2: Changes in observed global sea level since 1970, compared with the IPCC Fourth Assessment Report sea-level rise projections

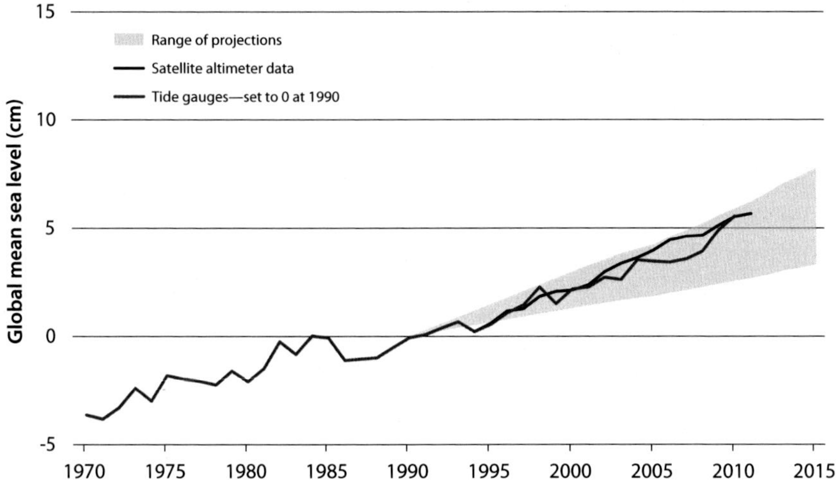

Note: Observational estimates of global averaged sea level estimated from tide gauges and the satellite altimeter data are shown in blue and black respectively. The shaded area shows the full range of global averaged projections of sea-level rise based on the scenarios used in the 2007 IPCC Fourth Assessment Report up to 2015. These projections do not include an additional allowance for a potential rapid loss of the Greenland and west Antarctic icesheets, which only becomes significant in the IPCC projections after about 2020. The tide gauge data is set to zero at the start of the IPCC projections in 1990 and the altimeter data is set equal to the tide gauge data at the start of the record in 1993.

Source: J.A. Church, J.M. Gregory, N.J. White, S. Platten and J.X. Mitrovica 2011, 'Understanding and projecting sea-level change', *Oceanography* 24(2): 84–97, updated from S. Rahmstorf, A. Cazenave et al. 2007, 'Recent climate observations compared to projections', *Science* 316(5825): 709.

The IPCC's Fourth Assessment Report in 2007 estimated that in a scenario similar to the 2008 Review's no-mitigation scenario, sea levels would rise in the range of 26 to 59 centimetres by 2100, with a lower limit for all IPCC scenarios of 18 centimetres. This figure did not include the potential dynamic losses from the Greenland and west Antarctic icesheets (increased calving), which could increase the upper end by about 10 to 20 centimetres by the end of this century. The IPCC also concluded that larger values above this upper estimate could not be excluded.

Quantitative estimates and upper limits for the contribution of the potentially rapid response of icesheets were not included because no consensus could be reached on the potential magnitude of these contributions by 2100.

The large land-based icesheets are currently losing mass to the ocean through both melting (Greenland) and dynamical flow (Greenland and Antarctica). The recent acceleration in the dynamical flow of both icesheets is thought to have been the result of incursion of relatively warmer ocean water underneath iceshelves. The warming leads to basal melting and thinning of the iceshelf, reducing the buttressing effect of the iceshelf on the icesheet, and the ice flow on land consequently accelerates towards the ocean, as observed in the Antarctic Peninsula.

There is considerable uncertainty whether this dynamical flow will continue as observed, or accelerate or decline in the future. The trends are based on shorter-term observation records and therefore are more difficult to distinguish from natural variability. However, a review of all observations shows that there is a net loss of mass from the Greenland and Antarctic icesheets. The uncertainty is about the rate at which this ice loss is occurring, not whether it is occurring at all. This uncertainty about the dynamics of the icesheets means that there could be a larger sea-level rise than current projections but not a significantly smaller rise.

There has been a significant focus on rates of sea-level rise and the future of the icesheets since the IPCC Fourth Assessment Report. The fact that observed sea-level rise is tracking near the upper limit of IPCC estimates has raised concerns that the IPCC projections may be underestimates. This is particularly so in the context of the current inability to adequately model the response of icesheets to global warming. These concerns have led to the development of several models of sea-level rise that use 20th century sea level and temperature records.

These 'semi-empirical' models all indicate larger rates of rise during the 21st century than the IPCC Fourth Assessment Report projections, with upper values as high as 1.9 metres by 2100. However, significant concerns have been raised about whether these projections are robust.

Other work suggests that a sea-level rise of more than two metres by 2100 is not physically possible, and that a more plausible rise—including icesheet contributions—is 80 centimetres, near the upper end of the IPCC estimates. The upper limit of sea-level rise in the 21st century is a matter for continuing research. There has been no credible publication of views that sea-level rise could be less than that suggested by the IPCC.

Changes to ecosystems

Another clear signal that warming is well under way can be found in the changing behaviours of various ecosystems. Research is showing that a warming signal is now evident in an increasing number of Australian and global observations of species. These include the southward expansion of the breeding range of black flying foxes and shifts in the timing of plant flowering. In some cases, such as the early emergence of butterflies in Melbourne, these changes have been attributed to climate change.

Responses to warming have also been observed in marine ecosystems, including the southward shift or extension of sea urchins and intertidal species. An important example is the increase in bleaching events on the Great Barrier Reef. There have been eight mass bleaching events on the Great Barrier Reef since 1979, with no known widespread bleaching events prior to that date.

Future climate change in Australia is likely to have impacts on ecosystems through increases in land temperatures and an increase in the variability, along with an overall decline, in rainfall in southern Australia. Major threats to ecosystems include extended drought periods, invasion of weeds and pests encouraged by the change in climate, altered fire regimes, land-use changes, direct temperature effects, increases in salinity and other water quality issues and changes in water availability.

Australia's biodiversity is not distributed evenly over the continent but is clustered in a small number of 'hotspots' with exceptionally rich biodiversity. Most of these areas, as well as many of Australia's most valued and iconic natural areas, are among the most vulnerable to future climate change. They include the Great Barrier and Ningaloo Reefs, south-west Western Australia, the Australian Alps, the Wet Tropics of Queensland and the Kakadu wetlands.

Predicting the future effects of climate change on Australia's biodiversity in these iconic areas and elsewhere is challenging for a variety of reasons. The effects of climate change will interact with other effects of human activities on biodiversity. Properties of ecological systems—communities of interacting species and their environment—are often complex, and can be difficult to understand and predict. A change in the average value of a variable, such as temperature, may not be as important ecologically as a change in the variability or extremes of that variable. Also, basic knowledge is generally lacking about limiting factors, genetics, dispersal rates and interactions among species that comprise Australian ecosystems.

Furthermore, many of the most important impacts of climate change on biodiversity will be the indirect ones, acting together with other factors. For example, for the Kakadu wetlands, the major threats of climate change are not the direct impacts on vulnerable species but rather an intersection of effects due to changing fire regimes, rising sea level and the resulting saltwater intrusion into freshwater wetlands, as well as the consequences of climate change for a suite of invasive weed and feral animal species.

Tipping points

Some climate change outcomes can be projected in a linear form based on current tendencies. There are also risks of abrupt, non-linear and irreversible changes in the climate system. These outcomes may have high consequences due to the extent or speed of the change. Other climate outcomes may have high consequences due to the numbers of people affected, including through the loss of ecosystem services such as pollination.

Some elements of the climate appear to be unresponsive to changes until a threshold is crossed, after which the response can be sudden, severe or irreversible. This threshold is referred to as a 'tipping point'. There is considerable uncertainty regarding the temperature at which this point may occur, as well as the likelihood of a given degree of human-driven climate change triggering any of these events.

One phenomenon that is irreversible is the acidification of oceans, which is caused by carbon dioxide dissolving in seawater. This has the potential to significantly affect marine organisms and ecosystems, including those that sustain important fisheries.

Measurements indicate that the average seawater acidity has increased by 30 per cent since pre-industrial times. Ocean acidification directly follows the accelerating trend in world carbon dioxide emissions, and the magnitude of ocean acidification can be determined with a high level of certainty based on the predictable marine carbonate chemistry reactions and cycles within the ocean. It is predicted that by 2050 ocean acidity could increase by 150 per cent. This is an added stressor for coral reefs because more acidic oceans lead to reduced calcification in corals.

There are a number of outcomes that could be considered extreme or high-consequence climate outcomes. They include changes to the El Niño – Southern Oscillation, the melting of the Himalayan glaciers, failure of the Indian Monsoon, the destruction of coral reefs and species extinction.

New research has focused on the tipping elements in the climate system. Progress has been made in identifying and testing potential early warning indicators of an approaching tipping point.

Attempts have also been made to better understand the probabilities of various tipping points by obtaining expert opinions from scientists. In a 2009 survey of 43 experts, each was asked about their views on the probability that certain major climate outcomes would occur. The results indicated that, while there is a range of views among experts about the prospect of major changes in the climate system being triggered, this does not necessarily imply that the probability of such outcomes occurring is considered to be low.

In fact, significant probability was allocated to some events, such as the dieback of the Amazon rainforest and melting of the Greenland icesheet.

The Amazon rainforest is the most widely cited example of a major plant and animal ecosystem at risk of abrupt change from a warming climate. Temperature increase, changes to the length of the dry season and the drought intensity anticipated under climate change will all influence the viability of the rainforest.

Simulations that incorporate the complex ecological processes in the rainforest system suggest that there is a threshold around a 2°C temperature increase above pre-industrial levels. Beyond that increase, the area of the Amazon forests subject to dieback rises rapidly, from 20 per cent to more than 60 per cent.

Severe droughts were recorded in the Amazon Basin in 2005 and 2010. The 2005 event was associated with the release of 5 billion tonnes of carbon dioxide due to the death and subsequent rotting of trees. Even larger emissions are expected as a result of the 2010 drought. Each time, the ability of the rainforest to absorb additional carbon dioxide is reduced.

Along with the observation that such droughts occur at the same time as peaks of fire activity, these recent events support the assessment that this ecosystem will be affected at relatively small temperature increases.

Such a dieback is an example of a carbon–climate feedback. These occur when changes in the climate affect the rate of absorption or release of carbon dioxide and other greenhouse gases from land and ocean sinks. Other examples of carbon–climate feedbacks include a reduction in the ability of the oceans to remove carbon dioxide from the atmosphere as water temperature increases, and the weakening of uptake by vegetation due to increased temperatures and reduced water availability.

There is also the risk of release of methane from permafrost and methane hydrate in the oceans as the world warms. This could lead to a positive feedback effect, where the increased temperatures cause a further release of these gases.

A recent study suggested that there are more than 1,700 billion tonnes of carbon stored in permafrost, which is about twice the amount stored in the atmosphere at present. It is unknown at what temperature this stored carbon might become unstable, or whether it would be released to the atmosphere over a short or long period of time. However, only about 100 of the 1,700 billion tonnes are considered to be vulnerable to thawing this century. Research on past and present emissions from these sources shows that current rates of emissions are low relative to overall global emissions, but it is not known whether these are new sources or just newly observed.

While the existence of tipping points can be anticipated with high confidence, specific thresholds at which they will occur cannot yet be predicted. Actively managing ecosystems to improve their resilience is important to ensure that the services that economies depend on—including pollination of crops and native vegetation, shade and shelter, maintenance of fertile soil and productive oceans, clean water and climate regulation—are available over the longer term.

Widespread ecological restoration could play an important role in ensuring the provision of ecosystem services and the maintenance of biodiversity. By addressing the range of pressures caused by human activities that, in combination, may push an ecosystem past a tipping point, we can help avoid or at least reduce the possibility of crossing a critical threshold. While climate change is a common driver of tipping point scenarios, in addition to reducing greenhouse gas emissions, investing in actions to improve ecosystem management will be needed to strengthen the ability of ecosystems to absorb and recover from shocks and reduce the risk of reaching irreversible tipping points.

Correlation is not causation

The dynamic and unpredictable nature of the earth's climate can make the detection of a climate change trend difficult. Even if observations are showing that trends are occurring in a range of climate variables, detection of a climate change trend is not the same as determining the cause. We need further evidence to establish a link between the observations and the cause.

The temperature of the earth and its atmosphere is determined by the balance of the incoming solar radiation and the heat that is radiated by the earth back into space. Temperature changes can occur as a result of more or less radiation coming in, or a change in the amount of outgoing radiation that is trapped by the atmosphere.

This balance can be influenced by a range of disturbances, including the sun's output, volcanic eruptions and, over hundreds of thousands of years, changes in the earth's orbit. To establish whether humans are responsible for the warming trend over the last 50 years, scientists need to establish that the changes are not explained by these natural factors.

Changes in the amount of solar radiation reaching the earth have been implicated in temperature fluctuations of the last 10,000 years. For the last 150 years, and especially since 1970, changes in solar output have been tracked with greater accuracy. Recent research suggests that solar output could have contributed at most 10 per cent to the observed warming trend in the 20th century, so other warming influences need to be considered.

Other important influences on the weather are shorter-term modes of natural variability, such as the El Niño – Southern Oscillation and the North Atlantic Oscillation. These phenomena may cause significant climatic variations on a year-to-year basis, but they cannot explain globally synchronous trends in temperature that occur from decade to decade.

To distinguish the contribution of greenhouse gases to observed trends from other potential influences, scientists have identified 'fingerprints of forcing'. These 'fingerprints' show patterns of change that are consistent with warming caused by greenhouse gases, rather than other sources, such as solar radiation.

One 'fingerprint' is the pattern of warming in the layers of the atmosphere. Models predict—and observations have confirmed—that the lowest layer of the atmosphere (the troposphere) is warming, while the next layer up (the stratosphere) is cooling. Increased output from the sun would be expected to warm both layers. This pattern can be explained by increases in greenhouse gases and the depletion of the ozone layer.

Scientists have also been able to use improved observational data to resolve what were viewed as inconsistencies between observations and expectations. Greenhouse theory and modelling anticipated that a hotspot should occur in the atmosphere about 10 to 15 kilometres above the earth's surface at the tropics, but this was not previously supported by observations. More accurate temperature observations are now available and greater warming has been detected in that area, which has provided another 'fingerprint' of changes caused by greenhouse gases.

Since the IPCC Fourth Assessment Report in 2007, climate model simulations have been run that reinforce earlier conclusions that both natural drivers (volcanic aerosols, solar variations and orbital variations) and human drivers (greenhouse gases and aerosols) are required to explain the observed recent hemispheric and global temperature variations. But greenhouse gas increases are the main cause of the warming over the past century.

Conclusion

In order to understand the mechanisms and implications of climate change, an interested non-scientist must draw on the publications of experts in the field. In this sense, the challenge facing each of us can be compared to that facing a judge in a court of law, who must make a decision on a balance of probabilities. How often does a case come before one of Australia's superior courts where the defence is so weak that it cannot find a so-called expert to blow a fog through the proceedings? The judge's job is to avoid wrong steps through the fog—to assess the chances that the opinion of just one so-called expert is more likely to be right than the established opinion.

The evidence for the prosecution in this case is considerable. The most important and straightforward of the quantitatively testable propositions from the mainstream science—upward trends in average temperatures and increases in sea levels—have been either confirmed or shown to be understated by the passing of time.

Some important parameters have been subject to better testing as measurement techniques have improved and numbers of observations increased. On these, too, the mainstream science's hypotheses have been confirmed. They include the warming of the troposphere relative to the stratosphere, and the long-term shift towards wet extremes and hot extremes.

The science's forecast of greater frequency of some extreme events and greater intensity of a wider range of extreme events is looking uncomfortably robust.

A number of measureable changes are pointing to more rapid movement towards climate tipping points than previously suggested by the mainstream science. The rates of reduction in Arctic sea ice and the accumulation of methane in the atmosphere are examples.

Indeed, scientific developments since 2008 have introduced additional caution about whether 'overshooting' emissions scenarios—where green-house gas concentrations peak above a goal before declining—will lead to temperature increases that are not quickly reversible.

Regrettably, there are no major propositions of the mainstream science from 2008 that have been weakened by the observational evidence or the improved understanding of climate processes over the past three years.

The politicisation of the science as many countries have moved towards stronger action to reduce greenhouse gas emissions has placed institutions conducting the science under great scrutiny. Exhaustive reviews have revealed some weaknesses in execution of the scientific mandate, but none that is material to the reliability of the main propositions of the mainstream science.

There is still a high degree of uncertainty about myriad important details of the impact of increased concentrations of greenhouse gases. But the uncertainty in the science is generally associated with the rate and magnitude, rather than the direction, of the conclusions.

Indeed, the consistency of the understatement since climate change became a large policy issue in the early 1990s is a cause for concern. It would be much more of a surprise if the next large assessment of the IPCC in 2014 led to a downward rather than upward revision of expectations of damage from unmitigated climate change.

This raises a question whether scientific research on climate change has a systematic tendency to understatement. It may be tempting to correct for this by giving more weight to the more concerned end of published research. This would be a mistake. In a highly contested and complex scientific matter with immense implications for public policy it is important to base policy on the established propositions of the science.

In drawing our judgment on the science, the evidence is now so strong that it is appropriate that we move beyond the civil court parameters of 'balance of probabilities' that I applied in 2008 towards the more rigorous criminal court conclusion of 'beyond reasonable doubt'.

2 Carbon after the Great Crash

ON THE MORNING of 30 September 2008, I handed *The Garnaut Climate Change Review* to the then Prime Minister of Australia, Kevin Rudd.

First, however, the Prime Minister wanted to talk about an urgent calamity. Overnight—that morning Australian time—the New York Stock Exchange had suffered its largest-ever points fall. All of the media talk was of the collapse of the international financial system and of imminent global recession. The collapse did indeed turn out to be great, with most of the main Wall Street financial institutions disappearing, or being taken over by others, and in any case being rescued by government. This massive restructuring of the centre of global finance was accompanied by a freeze in global markets the like of which had not been seen in 80 years. The crash was followed by an equally dizzying plunge in world trade as a series of global economic imbalances were corrected with ruthless speed.

In developed economies the Great Crash was followed by a Great Recession, the largest blow to growth since the Great Depression of the 1930s. Overall, the growth in economic output in developed countries contracted by 2.7 percentage points from 2007 to 2009, with the biggest falls in Japan and Europe. Australia was one of the few developed countries to avoid recession, with GDP growing by 3.6 per cent from 2007 to 2009.

However, as precipitous as it was, the Great Crash proved to be only a temporary deceleration of growth in developing countries. Led by China and India, but extending to and beyond Indonesia and Brazil, they returned quickly to the strong economic growth that had characterised the early 21st century.

The Great Crash accelerated an emerging shift in the global economy. The developed countries of the northern hemisphere now face lower long-term growth paths. That, in turn, has shifted their projected carbon emissions onto a lower trajectory.

On the other hand, developing countries are now the growth engines of the world economy and the trajectory of their carbon emissions in the absence of mitigation policies has shifted moderately upwards.

The Great Crash has had other legacies. Government spending in developed countries increased dramatically in response to falling growth and the massive support provided to ailing financial institutions. In many countries—including China, the United States, the United Kingdom, Germany, the Republic of Korea and Japan—stimulus spending increased support for energy-saving and low-emissions technologies.

Higher levels of public debt from recession and its aftermath will cost governments more to service and over time will require some combination of lower government spending and higher taxes. A wave of fiscal austerity has already begun to sweep across Europe and is being extended to the United States and other developed countries. This affects short- to medium-term growth prospects as well as the ability of governments to fund investments in infrastructure, health, education and technology.

The Great Crash of 2008 and its aftermath have revealed vulnerabilities in the American and European economies that had previously not been apparent to most observers. Many of the fundamental weaknesses—for example in banking and economic policy frameworks—remain.

The outlook is more positive for developing countries. Prospects are good for continued rapid growth in China and India, and there is strong growth momentum in many other developing countries. Current growth rates are higher even than the estimates embodied in the 2008 Review's projections, which were themselves much higher than those incorporated in forecasts and projections of international organisations.

This broadening of global economic growth into the large developing countries is the essence of what I call the Platinum Age. It arises from the adoption of the techniques and ideas of modern economic growth by the populous economies of Asia. The resulting period of higher growth has been characterised by openness to trade and investment, generally cautious fiscal and monetary policies, and high and rising rates of savings and investment.

Chinese growth since the reform era began in 1978 has consistently outperformed expectations, the more so in the early 21st century. Driven by increasing wages and more productive use of capital, growth will remain strong into the 2020s. Productivity growth through improved labour skills and technological improvement can support China's rapid growth until the late 2020s. By 2030, its average incomes will be more than half of those in the advanced industrial countries. By then the Chinese economy will be similar in size to those of the United States and the European Union combined.

In the longer term, India has even stronger growth prospects from a much lower current base. Three factors have been fundamental to the acceleration of its economic growth over the past two decades. Most basic of all, as in China, there has been steady opening of the economy to international trade and investment. Second, India continues to undergo a demographic transition that is favourable to growth, with the proportion of the population of working age set to gradually increase over the next two decades. If this increasing 'demographic dividend' is accompanied by continued improvements in the skills of the Indian workforce and by continuing economic reform, then rapid economic growth can be sustained for decades to come. India can power ahead long after Chinese growth has eased to the more gradual rates of a high-income country starting in the late 2020s. The third factor has been the large increase in national savings rates as incomes have increased. This has allowed investment rates to rise without increased risks of economic instability.

Indonesia, the world's third most populous developing country, also has strong and sustainable growth momentum. Indonesia managed to maintain solid growth near 5 per cent through the years of the Great Crash, and is now returning to stronger growth rates. Since the traumatic democratic transition of the final years of the 20th century, it has been building the institutions that are necessary for sustained strong growth within a democracy.

These three large developing economies (China, India and Indonesia) are all at stages in which growth is highly energy-intensive. All three happen to have considerable domestic endowments of coal. Emissions of greenhouse gases will grow rapidly in the absence of strong and effective mitigation policies. Their own business-as-usual growth in emissions will quickly absorb the atmosphere's limited remaining capacity to absorb greenhouse gases without high risks of dangerous climate change. In the remainder of this chapter we examine the likely growth in greenhouse gas emissions over the next couple of decades under business as usual, and then look at the remaining capacity of the atmosphere to absorb greenhouse gases without our running high risks of dangerous climate change. The two sets of data indicate the size and the urgency of the global mitigation task that we face.

More energy

So, what kind of energy use and emissions growth would emerge from a Platinum Age if there were no climate change or mitigation? Answering this question is an artificial exercise but provides essential perspective on the global mitigation challenge.

The task is artificial because established mitigation policies have already bent the trajectory of future emissions significantly downwards. Business-as-usual emissions in many developed countries and in the major developing countries, most notably China, are now a thing of the past. Regrettably, as we will see, they are not so obviously a thing of the past in Australia.

The task is also artificial in not taking into account the possibility of damage to growth as a result of climate change over the next two decades. As discussed in Chapter 1, the science tells us that the effects of higher atmospheric concentrations of greenhouse gases are experienced with a lag of several decades. The increase in global average temperatures that has already occurred is the result of changes in atmospheric concentrations that were in place several decades ago. The greenhouse gas concentrations that have accumulated so far in the Platinum Age are on top of earlier accumulations. They will be the main determinants of warming to the end of the 2030s. Substantial climate change over the next three decades is now 'built in' to the global climate system and is likely to be constraining economic growth by the 2030s.

Some of the effects of climate change that will probably be evident before 2030 include the increase in intensity of extreme weather events that, among other things, affect global food production and prices. In some countries, food price volatility and other manifestations of extreme events may affect the stability of political systems in ways that feed back negatively into economic growth.

Unfortunately, lower rates of growth associated with political disorder in individual states, or in the international system, are unlikely to help reductions in emissions. The effects of lower growth on emissions are likely to be greatly outweighed by lower priority and effectiveness of mitigation policies.

To project business-as-usual emissions we must start with analysis of economic output. The projections ignore the effects of climate change on economic activity. The projections of output take into account trends in population growth, investment and the productivity with which labour and capital are used. Then judgments are made as to the likely energy intensity of this projected growth in output from which overall demand for energy can be derived. Finally, judgments are applied to the emissions intensity of energy demand which allow us to derive projections of carbon dioxide emissions. Changes in the emissions intensity of energy demand will reflect different sources of energy supply, developments in energy efficiency and related technological developments.

The chain linking economic output to energy demand and energy demand to carbon emissions is shown in Figure 2.1.

Energy intensity is a measure of how much energy is used per unit of economic output. It has declined over time in most countries as more energy-efficient equipment is used and as a greater share of economic activity comes from services and other less energy-intensive activities. There are, however, great differences in the underlying rate of change in different countries, depending on their stage of development, resource endowment, economic structure and other factors.

Figure 2.1: The decomposition of emissions growth

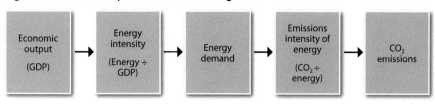

The outlook for future improvements in energy intensity is also affected by government policy, as well as by changes in the expected future costs of energy.

Looking forward to 2030, energy intensity of output in the absence of mitigation policies is projected to fall in all regions, with the ratio of energy consumption to GDP declining at average rates of between 0.9 and 2.7 per cent per year. The global average is projected to decline by 1.9 per cent annually between 2005 and 2030. Expectations of substantially higher fossil fuel prices have an important effect.

Applying these projections of energy intensity to the projections of economic growth yields projections of total energy use. These indicate that growth is tailing off across developed countries to a 0.2 per cent annual increase over the period 2005 to 2030.

Developing countries, by contrast, are projected to experience continued strong growth in energy use, at 4.7 per cent per year. Underlying energy demand growth is fastest in the rapidly growing economies of Asia, but strong across most developing countries.

More emissions

The extent to which increases in energy demand translate into increased emissions depends on the amount of emissions per unit of energy consumed (or the carbon intensity of energy). Carbon intensity is largely determined by the fuel mix in the energy system. Among the fossil fuels in common use, coal is the most carbon-intensive energy source, followed by oil, then gas. The energy mix changes over time, including varying proportions of nuclear and renewable energy that produce close to zero carbon emissions.

The projections under business as usual are for a 0.3 per cent per year increase in the global carbon intensity of energy supply between 2005 and 2030. There are variations between countries in a relatively narrow band. Carbon intensity is projected to fall slightly in developed countries, and to increase on average in the developing world. This is in line with recent trends, but will moderate over time. India is projected to experience the fastest rate of increase in carbon intensity, because so much of its energy supply is derived from coal. China's carbon intensity rose significantly from 2005 to 2009 as the importance of coal in the mix increased, but under these projections remains constant over the next two decades.

The prevailing fuel mix differs greatly between countries, depending on the availability and cost of different energy sources. The nature of the electricity supply system, the structure and size of energy-intensive industries, as well as of the transport and housing stock, also play a role.

Changes in the fuel mix tend to be relatively slow, as they mainly come about through gradual replacement of a large stock of long-lived energy infrastructure. A country's long-term fuel mix, in practice, is strongly influenced by climate change policies as well as energy policies—for example, energy policies directed at reduced reliance on overseas purchases.

Relative prices of different energy sources are the other principal determinant of the fuel mix. If oil prices increase relative to prices of other energy sources, as is widely considered to be likely, this will trigger a substitution away from oil and towards other energy sources. The effect on carbon emissions is unpredictable. Replacement energy sources may be lower in carbon intensity, as in the case of electricity from gas, nuclear or renewable sources. Alternatively, they may have greater carbon intensity, as in the case of coal, and coal-based liquid fuels, tar sands and conversion of gas to liquids.

For the time being, the price of gas relative to oil will tend to fall in the United States and elsewhere as new supplies emerge in high volumes. However, the International Energy Agency argues that the decoupling of contract gas prices from oil prices does not necessarily mean weaker gas prices in the longer term. The price incentives to replace coal and oil by gas are likely to be positive and influential for a considerable while, but at some time will come under pressure from perceptions of scarcity.

A comparison of the global energy mix at 2030 under old and new International Energy Agency projections illustrates the point. The agency's most recent projections have the share of oil in global energy supply at 2030 at 28.5 per cent, down from 31.5 per cent in previous projections. This reduction is made up in roughly equal measure by increased use of coal, nuclear power and electricity from renewables, including biomass. The substitution effects differ greatly between countries, but the net effect of moving from the old to the new projections is to slightly reduce the expected carbon intensity of the global energy supply.

Amalgamating the above projections of the individual drivers of business-as-usual emissions, we find that global emissions to 2030 rise at an average of 2.8 per cent per year. Developed countries contribute 30 per cent of global emissions in 2030 under a business-as-usual scenario. Developing countries contribute 70 per cent of global business-as-usual emissions at 2030, up from 50 per cent today. China's and India's share in global emissions would be 41 and 11 per cent respectively. The share of other developing countries would remain at 19 per cent of the global total.

Total global emissions from fossil fuels are projected to double between 2005 and 2030 under business as usual. This is a similar perspective to that presented in the 2008 Review. The division of growth in emissions between developed and developing countries, however, changes dramatically. Developed country emissions are now expected to fall slightly between 2005 and 2030 under business as usual. Developing country emissions are expected to rise a bit more rapidly than anticipated in 2008. Total world emissions growth is expected to be slightly lower than was expected by the 2008 Review. The total burden is similar to that of three years ago, but more of it relates to emissions in developing countries (see Figure 2.2).

Figure 2.2: Projections of average annual growth in emissions in the absence of mitigation policies, 2008 Review and 2011 Update, 2005 to 2030

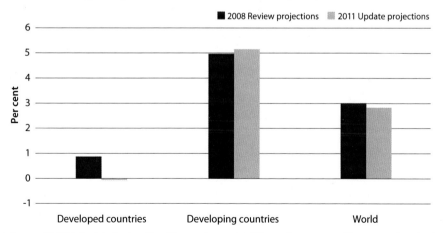

Sources: BP 2010, *Statistical review of world energy*; International Energy Agency 2010, CO_2 *emissions from fossil fuel combustion 2010* and *World energy outlook 2010*; 2008 Review.

Australia's emissions under existing policies

Australia weathered the Great Crash of 2008 better than any of the major advanced countries.

The factors that have driven the strong growth in Australia's economy and energy demand can be expected to continue for at least several years. Even with the mitigation policies that were in place by late 2010, emissions would grow strongly in the period immediately ahead as a once-in-history resources boom reaches its greatest height. In this, Australia is different from other developed countries. Australia's natural resources are likely to continue to provide fuel for the growth in China and India in the years ahead. The relative importance of gas, coal and uranium is likely to change over time in ways that depend on technological and price developments as well as climate change mitigation policies.

The strength of the resources sector will encourage Australian emissions growth through two channels. The first is through the effect of increased economic activity on energy demand. The second is through the emissions that are a by-product of the extraction of natural gas and coal.

In the year to August 2010, gas generation supplied more than 11 per cent of eastern states' electricity demand, while black coal supplied 56 per cent and brown coal 24 per cent. Renewables, including hydroelectric and wind, supplied the remaining 9 per cent.

Australia releases annual projections of our emissions on the basis of current policies. The 2010 report shows that Australia is on track to meet its Kyoto Protocol target of limiting emissions to an average of 108 per cent of 1990 levels between 2008 and 2012, with emissions projected to reach 106 per cent of 1990 levels (see Figure 2.3).

However, the report forecasts strong growth in emissions in the absence of further policy action. Emissions are projected to increase by 24 per cent from 2000 levels by 2020. This represents a 4 per cent upward revision of the previous year's projections, and is 4 per cent above expectations in 2007.

Australia's 2020 emissions target, as reported to the United Nations Framework Convention on Climate Change, is an unconditional 5 per cent reduction relative to 2000, with conditional targets extending to a reduction of 25 per cent depending on the actions of others. The projected growth in emissions thus presents a substantial mitigation task in the decade ahead, and obviously higher still with higher levels of ambition.

Figure 2.3: Australia's emissions trends, 1990 to 2020

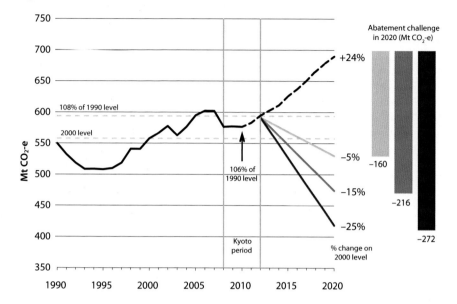

Note: Mt CO₂-e is megatonnes of carbon dioxide equivalent; 'abatement challenge' means the reduction in greenhouse gas emissions required to reach specific targets; and the bars represent the levels of abatement (in Mt CO₂-e) that would be required to reach the three targets for 2020: minus 5 per cent, minus 15 per cent and minus 25 per cent of 2000 levels.

Source: Australian Government 2010, *Australia's emissions projections 2010*, Department of Climate Change and Energy Efficiency, p. 8.

The strong growth in emissions over the coming decade is expected to be dominated by the extraction and preparation of energy resources for export. Over a third of the increase in emissions growth to 2020 derives from increased fugitive emissions from growing coal and gas exports. The total of fugitive emissions from coal mining and oil and gas extraction plus the direct fuel combustion emissions from gas liquefaction projects accounts for almost half the growth in Australia's emissions to 2020. Electricity generation is expected to play a much smaller role in emissions growth in the coming decade than in the past.

Budgeting carbon

If we juxtapose this assessment of business-as-usual emissions with the requirements of avoiding dangerous climate change, we get a good sense of the challenge.

The 2008 Review recommended, and the Australian Government and Opposition accepted, that it was in Australia's national interest to seek a global emissions concentrations objective of 450 parts per million (ppm) of carbon dioxide equivalent. This is broadly equivalent to the objective of holding temperature increases to 2°C above pre-industrial levels, which was agreed by the international community at the Cancun meeting.

To hold concentrations at a level that would mean warming of no more than 2°C would require a major elevation of the importance of climate change in national priorities.

From the perspective of 2011, for the world to hold emissions concentrations to 550 ppm carbon dioxide equivalent would be an achievement of international cooperation and innovation in national economic policy of large dimensions. To achieve 450 ppm carbon dioxide equivalent with only the degree of overshooting envisaged in the 2008 Review would be an achievement in international relations and national public policy of historic dimensions. The path to anything lower than 450 ppm carbon dioxide equivalent now has to involve overshooting.

In this light, the most important development in scientific discussion of mitigation goals since 2008 is an increasing focus on a cumulative emissions budget. Such a budget approach was favoured conceptually in the 2008 Review. The Review formulated the global mitigation problem as one of optimal depletion of a finite resource. The resource in this case is the earth's capacity to absorb greenhouse gases without dangerous climate change.

Cumulative carbon dioxide emissions can be determined so that a 'budget' can be defined that is essentially independent of timescale and trajectory.

However, it is not possible to achieve this for the full set of greenhouse gases over long time periods. For gases with a lifetime shorter than a few decades, the rate of emissions at a particular time has a strong influence on concentrations, and hence impacts, at that time. Similarly, longer-lived gases such as carbon dioxide and nitrous oxide can be expected to have a larger influence in the longer term; with carbon dioxide having by far the largest influence.

The main value of the cumulative, or budget, approach is to focus attention on the limited volume of greenhouse gases that can be released into the atmosphere over specified periods without creating large risks of dangerous climate change. The basic arithmetic within this approach is sobering. One recent study analysed the allowable global cumulative carbon dioxide emissions between 2000 and 2050 in terms of a number of different probabilities of exceeding 2°C of warming. The study found that a budget of 1,440 gigatonnes would have a 50 per cent chance of holding temperature increases to 2°C. An estimated 350 gigatonnes of carbon dioxide were emitted globally between 2000 and 2009, which represents about one-quarter of the total budget for 2000–50.

So at the rate of emissions of the first nine years of the century, the remainder of the budget would be exhausted by 2045. The world's emissions are already well above the average for 2000–09, and rising strongly. A major change of trajectory is needed quickly if 2°C is to remain a realistic possibility.

Overshooting

Overshooting scenarios were first considered around 2004, when it was recognised that such an approach would be necessary if a decision were made to aim for stabilisation of greenhouse gas concentrations at, or close to, those that would constrain temperature increases to 2°C.

Overshooting scenarios allow a slower initial reduction in emissions. Some models have shown that the slow response of the climate system can allow a small, short overshoot in concentration without a corresponding overshoot in temperature. However, for a given concentration stabilisation target, any amount of overshoot increases the risk of reaching a level of climate change that could be considered dangerous.

To achieve reductions in atmospheric concentration and eventual stabilisation, emissions must fall below the natural level of removal from the atmosphere by the oceans and biosphere. As we have seen, the rate of removal can be affected by climate change itself, an outcome referred to as a 'carbon–climate feedback'. Research suggests that the rate of uptake by ocean and land sinks decreases as higher temperatures and greenhouse gas concentrations are reached.

The major risk and uncertainty associated with overshooting is the level of climate change reversibility. Some models suggest that it may be possible to reduce atmospheric concentrations of greenhouse gases significantly over one or two centuries. Other models indicate that the rate of reduction in concentrations may be considerably slower, and there is also the chance that the climate may be pushed past a point of no return.

While the timing of the climate response is still uncertain, an overshooting scenario is more likely to lead to containment of temperature increases than a scenario where concentrations are stabilised and held at the 'peak' level.

It is becoming more difficult for a concentration overshoot to be 'small and short' with an objective of 450 ppm carbon dioxide equivalent or 2°C. While ambitious greenhouse gas concentration objectives are becoming increasingly reliant on an overshooting or peaking scenario, new science suggests that, when compared to stabilising concentrations without overshooting, such a scenario may be less successful in avoiding long-term temperature increases than previously expected.

If we are to avert serious overshooting to achieve the 2°C target, dramatic changes in emissions trajectories are required in the next decade.

It may be that much deeper emissions cuts in many countries will become possible following a large global political response to the reality of worsening impacts of climate change. That is, a catastrophe may result in a strong political response. However, at that point, the lags between emissions and warming would have locked in a good deal of additional warming.

The difficulty of attaining the 2°C goal increases as the momentum of emissions gathers with the expansion of modern economic growth in developing countries and with the slow start to large emissions reductions in developed countries.

Recognition of this reality has increased the attraction of approaches to mitigation that may delay for a while the full impact of warming from greenhouse gases, and of technologies that have the capacity to remove emissions from the atmosphere.

Solar radiation management—the placement in the high atmosphere of particles that deflect solar radiation away from the earth—has come into closer focus as a temporary buffer to the immediate consequences of global warming, while major emissions reductions are implemented. Recently this and other geoengineering techniques have moved from science fiction into the realm of responses that are being subject to analysis. And while geoengineering has the potential to help mitigate greenhouse gas emissions, in the absence of regulatory mechanisms, there are concerns about possible negative consequences.

A recent report looking at black carbon and tropospheric ozone noted that reducing these pollutants and their precursors, which have a relatively short lifetime in the atmosphere, would slow the rate of climate change within the first half of this century.

The reductions in near-term warming could be achieved through the recovery of methane from fossil fuel extraction and transport, methane capture in waste management, use of clean-burning stoves for residential cooking, use of diesel particulate filters for vehicles and the banning of field burning of agricultural waste.

Actions to reduce black carbon and solar radiation management would complement but not replace reductions in long-lived greenhouse gas emissions. Reductions would still be required to protect the climate in the long term, and resolve issues such as increased ocean acidification.

The urgency also increases the importance of sequestration—the capturing of carbon in geological or biological sinks—to keep alive the possibility of limiting global average temperature increases to 2°C above pre-industrial levels.

Chemical processes, biological processes and carbon capture and storage have all been suggested as possible methods for the sequestration of carbon. Carbon dioxide removal through biological processes, particularly reforestation and algal sequestration, are generally more mature technologies than solar radiation management, with fewer uncertainties. They have a good track record over many hundreds of millions of years. Their role in sequestration could be extended if accompanied by the harvesting and storage of hydrocarbons.

Conclusion

The shift in the centre of gravity of global growth towards the developing countries makes the mitigation effort more difficult. Even by 2030, the average income per person in developing countries would be only a little over a quarter of that in the United States. Today's differential is well over twice that large, so the next two decades would see a substantial narrowing of the gap in economic activity and living standards, but nowhere near its closure. China would see its per person income at above half that of the United States in purchasing power parity terms. In India, purchasing power per person would remain below a quarter of that in the United States.

The same continued potential for catch-up exists for greenhouse gas emissions. Seventy per cent of global emissions by 2030 would come from developing economies that are home to 80 per cent of the global population. Under business as usual, China's annual emissions per person would reach those in the United States by 2030 (at around 15 tonnes of carbon dioxide per person from fossil fuel combustion only). Across the developing countries, however, emissions per person would still only amount to 38 per cent of those in the United States (or China), with obvious strong potential for further growth.

However, there is also considerable hope. The picture changes dramatically under effective mitigation. As we will see in the next chapter, the developing world and China in particular have already moved a long way from business as usual. Developing countries have many exceptional opportunities to de-carbonise their energy supply, to increase energy efficiency, and to cut emissions through practices such as better forestry management, agriculture and industrial activities. It is often less costly not to enter a path of carbon-intensive development than to disentangle an economy from an established carbon-intensive structure. These circumstances of developing countries will need to be brought fully to account if the reasonable ambitions for improved living conditions of the majority of humanity are to be reconciled with the avoidance of dangerous climate change.

One important driver of emissions—global population growth—has been gradually easing over the past several decades in response to rising living standards in the developing world—reinforced or compounded in China by strong anti-natal policies. The acceleration of economic growth in developing countries in the early 21st century holds the prospect for further reductions in fertility and population growth.

Second, a surprising expansion of global gas reserves, in the United States and in many other countries, creates an opportunity for greater reductions in emissions intensity of energy use than anticipated in the business-as-usual projections, through gas replacing more emissions-intensive coal.

For the developed countries, the decline in expected emissions growth following the Great Crash makes it easier to realise announced mitigation targets. But this easing of demands on emissions in developed countries is matched by increased demands from developing countries. The arithmetic of greater concentration of anticipated global emissions growth in the developing countries points to the need for earlier and stronger constraints on emissions in developed and developing countries alike.

The emissions growth outlook is especially challenging for Australia. We stand out as the developed country whose anticipated business-as-usual emissions growth bucks the general trend of developed countries, largely as a result of the expansion of the relative role of resources in the economy.

Existing policies leave exceptionally high anticipated growth in emissions to 2020. This will not be easily understood by other countries, and is likely to bring Australian mitigation policy under close scrutiny. Any failure by Australia to do what others see as our fair share will invite critical and perhaps damaging international responses, as well as damaging some other countries' commitments to mitigation.

3   What's a fair share?

WHEN HE was mayor of Shanghai in the 1980s, Jiang Zemin discussed Chinese–Australian relations with me on many occasions. Shanghai is a great centre for steel and wool textiles and he recognised the quality of Australian services, so we had a lot to talk about. Over dinner we would move on to the respective contributions of American presidents to their country and to the world. When I joined Jiang in reciting the Gettysburg Address with the fruit at the end of a meal, he kept us going when a word slipped my mind.

Back in Beijing, at the beginning of the 1990s, China's recently appointed president invited me to call in. Democracy sounded good in theory, he told me, but it didn't work in practice. You had to be rich to reach the top of the political system, he said. Money bought policy. In China under the leadership of the Communist Party, government could make decisions in the national interest and enforce them in the national interest.

I responded that it wasn't at all like that in Australia. I explained the autonomy of Australian political leaders, and the role of independent perceptions of the national interest in big Australian policy decisions.

Twenty years later, a different Chinese president and the premier met Prime Minister Gillard on a Chinese visit in April 2011. Jiang would have received a report from all of the meetings of the Australian delegation, and I hope that he saw the evidence on my side of our old argument.

The visit was joined by senior Australian business leaders, who met with the leaders of major Chinese state-owned enterprises. The Chinese delegation was led by Chen Yuan, chairman of the China Development Bank and son of a legendary Communist Party figure from the great debates of the 1950s through the 1980s, Chen Yun. The Australian delegation comprised a 'who's who' of Australian big business.

Chinese business leaders reinforced the Chinese Government's emphasis on energy efficiency and reducing emissions in their country's 12th five-year plan. Australian business figures raised doubts about the Australian Government's carbon tax.

One wonders what Jiang has made of it all.

Over the past four years, China has moved decisively on policy and action on climate change. Four years ago China was hiding behind the Kyoto special treatment of developing countries to resist making any commitments at all. Today, the transition to a low-carbon economy is a central feature of the five-year plan. China is shifting the relationship between economic growth and greenhouse gas emissions.

The Australian political community, on the other hand, was just deciding to catch up with the rest of the world in May 2007, when the Howard government received the report of the Prime Ministerial Task Group on Emissions Trading, but is more undecided today, despite Australia being the developed nation that is expected to be most badly affected by unmitigated climate change.

The 2008 Review argued that only a world bound by agreement on greenhouse gas reduction could avoid great damage from climate change. Acting on this proposition, however, entailed resolving a 'prisoner's dilemma'— a situation in which each country pursuing its own narrow self-interest would make decisions whose overall effect would be the worst possible outcome for them all. I described this then as a diabolical policy challenge.

Remarkably, the world is resolving the dilemma. This breaks several expectations. From its inception in 1990, the United Nations process that was crystallised in the Kyoto Protocol in 1997 divided developed from developing countries, and only the developed were bound by a specific emissions target. The Kyoto agreement set a goal for developed countries together to reduce emissions by 5 per cent below 1990 levels between 2008 and 2012 (the first commitment period).

The position that only developed countries should have emissions targets was thought to be appropriate for four reasons: the developed countries were responsible for most of the emissions already accumulated in the atmosphere; they were still responsible for most emissions; they could more easily afford the cost of adjustment to a low-carbon economy; and they were in a better position to develop and deploy new low-emissions technologies. The Howard government spoke for us in 1997 and concurred.

In making major mitigation efforts, developing economies have overlooked the agreed ethical obligation of developed nations and overturned a decade or more of global diplomacy.

Kyoto revisited

So, how did this come to pass? Apart from the position that only developed countries should accept emissions constraints, the Kyoto model had other flaws. Rather than articulating principles for allocating responsibility for emissions responsibility, it left the job to politics, arm-twisting and negotiations.

The Kyoto agreement was also damaged by the refusal of two of the developed countries to ratify the agreement they had helped to negotiate.

The Clinton administration had not been able to secure congressional support for the ratification of Kyoto. The Bush administration that was elected in November 2000 elevated non-ratification to a policy objective. After the 2000 election in the United States, Australia followed a similar path until it ratified the Kyoto Protocol in 2007 in Bali.

When countries that had ratified the Kyoto Protocol met in Bali in December 2007 to continue discussions about post-2012 mitigation targets, three emerging realities shaped the agenda. The first was a universal recognition that human-induced climate change was 'unequivocal' and its fallout was going to be large. Second, there was a growing domestic interest in the United States from outside the administration and international pressure on the United States to commit its fair share to global mitigation efforts. Third, there was growing recognition that emissions from developing countries would account for much of the future growth in global emissions. Developing countries would have to constrain emissions sooner rather than later.

In response to these three challenges, parties to the UN Framework Convention on Climate Change reached an agreement on the Bali Action Plan, which set out a two-year road map for negotiations on two tracks to settle the scale and scope of post-2012 commitments. Developed countries were to agree on a new set of commitments on emissions reductions under the second commitment period of the Kyoto Protocol due to commence in 2013. All countries were to agree to a 'shared vision' in Copenhagen to underpin a post-2012 agreement consistent with a pathway to avoiding dangerous climate change.

In practice, the United States agreed to take on mitigation commitments or actions comparable to those of the other developed countries in the Kyoto Protocol. China and the developing world also agreed, for the first time, to consider mitigation actions. However, there was no agreement on whether the Bali Action Plan would lead to legally binding commitments by either developed or developing countries under the Convention.

It was clear well in advance of the climate change conference in Copenhagen in December 2009 that officials would be unable to deliver a clearly defined and comprehensive set of binding commitments to mitigation. It was this long-anticipated failure of the formal negotiations, as well as the diplomatic fiasco that accompanied it, that dominated media coverage of the event.

In the event, a group of global leaders pulled together what became known as the Copenhagen Accord. This was not formally agreed at Copenhagen. Since important elements of it were accepted at Cancun a year later, it has been recognised that the accord was a major and positive step in international climate change efforts.

The Copenhagen conference also marked the serious multilateral re-engagement of the United States on international climate change efforts after a gap of nine years. And on the developing country side, some major countries (Brazil, South Africa, India and China) emerged as an influential negotiating group. As well, interaction between China and the United States was, for the first time, explicitly at the centre of possibilities on a major multilateral issue.

The parties had stopped short of formally approving the Copenhagen Accord, thus preventing it from becoming the core decision of the Copenhagen conference. However, it was noted by the conference, and parties were invited to bring forward pledges to reduce emissions.

Following Copenhagen, the UN Framework Convention parties met again in Cancun in December 2010. In contrast to the exceedingly high expectations for Copenhagen, public expectations were low for Cancun. As it turned out, the 2010 meeting worked diplomatically and the Cancun Agreements consolidated and cemented the Copenhagen Accord. They are now widely seen as a new beginning for international climate change efforts. Cancun provides further building blocks towards a comprehensive international agreement that includes emissions constraints by all major emitters.

The main outcomes of Cancun were:

- agreement to anchor under the UN Framework Convention the content of the Copenhagen Accord, including:
  - a global mitigation target—that global warming should be limited to below 2°C above pre-industrial average global temperatures, with periodic review to consider strengthening this long-term goal, including to 1.5°C (the first review is to begin in 2013 and conclude by 2015)
  - pledges made by all developed and major developing countries to constrain emissions
- establishment of a new Green Climate Fund to support developing countries' climate change responses; a collective commitment by developed countries to provide US$30 billion in fast-start finance in 2010–12; and a commitment to mobilising US$100 billion per year in public and private finance by 2020 in the context of meaningful and transparent mitigation
- a mechanism to deliver economic opportunities for developing countries to reduce emissions from deforestation and forest degradation

- new rules on measurement, verification and reporting and international consultation and analysis to ensure that all countries can see what others are doing to tackle climate change
- agreement to provide strong and practical support for vulnerable developing countries to manage unavoidable climate impacts, including the establishment of the Cancun Adaptation Framework to enhance adaptation efforts by all countries
- establishment of a mechanism to help deploy clean energy technologies around the world.

Another important development at Cancun was Japan's unequivocal statement reiterating its position from Copenhagen—that, while it would commit to major reductions in emissions, it would not enter a target in a second Kyoto commitment period. Other countries, including Canada and Russia, are likely to follow Japan.

Cancun may therefore mark the beginning of the end of the Kyoto regime and, accordingly, the end of the twofold structure of climate change effort. This is not a bad thing, so long as a number of unequivocally positive achievements of the Kyoto years are preserved. An arrangement within which all countries make commitments to limit emissions and to report on their progress under one universal instrument is more likely to lead to an effective global outcome than the old separation of developed and developing countries.

The new regime

It is taking some time for the reality to sink in, but there was a fundamental change in the international climate change regime at Copenhagen and Cancun. The regime changed to meet the requirements of the United States and the major developing countries. The changes for developing countries were essential because of the new power relations that were emerging in the Platinum Age and accelerated by the Great Crash of 2008.

While the change in the United States presidency from Bush to Obama meant that the US Government was now a strong supporter of national and international mitigation, and the 2008 changes in the Congress were supportive of the president, the United States continued to baulk at being bound by an international agreement.

The Copenhagen Accord was worked out by the large countries that wanted, or whose national policies and objectives required, a change in the global regime—the United States, China, India, Brazil and South Africa. There was a strong global agreement at Cancun, but not the agreement that Australia, among a number of developed countries, had been working towards.

The new regime carries on a number of elements of the old. It continues the definition of greenhouse gases and the way of calculating national emissions, and the processes for agreeing to change these methods over time. It carries forward agreements and discussions of the several mechanisms for assisting developing countries with reducing their emissions and adapting to climate change. It defines emissions reduction targets in terms of emissions within one's own boundaries—a production rather than consumption basis for calculating national emissions. It allows countries to meet part of their emissions reductions requirements by purchasing entitlements from others who are overachieving on their own targets. Finally, it carries forward and in a sense fulfils several of the objectives of the Kyoto arrangements, most importantly through the objective of containing global warming to below 2°C.

It also changes or abandons some features and aspirations of the Kyoto regime. It has secured commitments to emissions constraints that are far wider in their scope and quantitatively far more important than those made at Kyoto in 1997. However, these have been voluntarily offered by the country making the undertaking, and have not been agreed in a political negotiation. The process, which had been suggested by the Australian delegation as a way out of an impasse in Copenhagen, is described as 'pledge and review', and it is hoped that the review component will lead over time to some broad equivalence among the pledges being made. This has been described as a 'bottom–up' approach, which is contrasted with the 'top–down' approach at Kyoto. And it does not pretend that these are legally binding agreements; they are serious national and international political commitments but are not enforceable in courts of law.

How important are the two main innovations in the Cancun form of international agreement: the unilateral nature of the emissions reduction commitments, and their voluntary—but legally unenforceable—nature?

There would have been advantages in a comprehensive global agreement on emissions entitlements covering developing as well as major developed countries that 'added up' to the global temperature objective and to the emissions budget that was implicit in that objective. We would have arrived at a set of national commitments that would, if implemented, solve the problem. And we would have had a firm basis for international trade in entitlements, where countries that were having difficulty meeting their targets purchased permits from countries that were able to constrain emissions by more than was required by the agreement.

The negotiated commitments on emissions constraints would have resolved the prisoner's dilemma problem in one hit, by assuring each country that it could rely on others doing enough to solve the global problem if it took strong action itself.

What, then, are the advantages of binding commitments? They would provide support for global international trade in entitlements. Trade would reduce the costs of emissions reductions for the world as a whole and for each country. Trade would also establish a single price for entitlements in all countries that participated without limits in trade in credible emissions permits. The firm and verifiable nature of the commitments, with international trade in entitlements, would give rise to the single price—the price would be the same in a country with targets that required large absolute reductions in entitlements as in one in which entitlements were based on reductions in emissions intensity of output. The single price, in turn, would remove distortions in international trade in goods and services associated with variations in the costs of reducing emissions across countries. It would therefore remove any case for assistance to trade-exposed industries in any country, with potentially large and beneficial implications for the public finances and for the integrity of policy-making processes in all countries.

These advantages were discussed at length in the 2008 Review. The advantages of a comprehensive and binding international agreement that adds up to the global emissions budget are as strong in 2011 as they were three years ago. Such an agreement remains the Holy Grail.

As we have seen, long before the Copenhagen conference in December 2009 it was clear that a 'top–down' agreement was beyond reach. Apart from anything else, it was clear by early 2009 that the work at the official level was simply not being done to allow heads of government to reach comprehensive agreement in anything but the most general terms.

A formal comprehensive global agreement on emissions entitlements remains beyond reach for the foreseeable future. Only an agreement built around non-binding, unilateral commitments, albeit disciplined by peer review, had—and has—any early prospect of being acceptable to the United States and to the major developing countries.

On the other hand, experience since Copenhagen has demonstrated that Cancun-style agreements, beyond being necessary to secure commitments from some countries that are crucial to a successful global emissions reduction effort, have the large advantage that they encourage greater ambition in each country's emissions reduction effort.

Some countries would be unable or unwilling to make any commitment at all if it were binding under international law. Others would be less ambitious. The United States and India at present fall within the first category, and China the second.

This tendency for international commitments to be stronger if they are not legally binding is not confined to climate change negotiations. Governments have often been prepared to go further with domestic trade liberalisation in the context of less formal discussions than in formal negotiations directed at a legally binding agreement. The far-reaching trade liberalisation right through the western Pacific region from the mid-1980s to the late 1990s occurred within a context and doctrine of 'concerted unilateralism', in which countries reduced their own trade barriers without formal reciprocity from others, but in confidence that they would not be entirely alone. Reciprocity was diffuse rather than specific. The pace of liberalisation everywhere slowed sharply when 'concerted unilateralism' gave way to formal negotiation of legally binding preferential trade agreements from the beginning of the 21st century.

The problem with unilateral non-binding commitments, at least in their early iterations, is that they are unlikely to add up to the required amount of emissions reductions. This is the case with the first round of commitments within the Cancun Agreements.

The other problem with non-binding commitments is that they provide a less firm foundation for international trade in entitlements.

Whether these are serious or even fatal flaws depends on what happens next.

What next for the international climate change regime?

Now that a set of commitments has been placed on the table, we can add them up and assess the extent of the global constraint on emissions that they imply. Some of the commitments, including the European and Australian, contain unconditional and conditional elements, so that they are extended if others are pledging more. Future pledges will be more ambitious and move more closely towards what is required to reach the global temperature goal if there is confidence that others are going forward as well. There is a possibility that we will move towards more ambitious goals through an iterative process. A possibility but, of course, not a certainty. What is certain is that we would not have got far in 2009 and 2010 if we had taken the view that we had to negotiate an agreement that added up to the Cancun 2°C objective in a single step.

In addition to developments within the UN Framework Convention, the international climate change regime is also likely to be influenced by discussions occurring outside these processes, including in the G8, G20 and Major Economies Forum. For example, since late 2009, fossil-fuel subsidy reform has been elevated in the international energy, climate change and finance agendas. In his opening statement to a conference hosted by the Global Subsidies Initiative of the International Institute for Sustainable Development and United Nations Environment Programme in October 2010, the Deputy Director-General of the World Trade Organization, Harsha V. Singh, characterised fossil-fuel subsidy reform as one of the most important tools to combat climate change.

Discussions relevant to the international climate change regime will probably continue to evolve for some time along a number of tracks. The benefits of this include greater collaboration and trust between countries, leading to a greater willingness to make deep, reciprocal commitments. These developments could then be brought to account at a later date in the context of the UN Framework Convention.

It turns out that the difference between binding and non-binding agreements is not as wide in practice as this description suggests. The notionally binding commitments under the Kyoto Protocol turned out not to be legally enforceable in practice; and all countries at this stage are treating Cancun statements on emissions constraints as serious domestic and international political commitments. The US Government is treating its Cancun target as a serious objective of domestic policy and taking steps towards its fulfilment. China and India have made achievement of their emissions intensity targets central features of their five-year economic plans.

To the extent that the commitments are objectively determined in specific and verifiable emissions targets, even if these are intensity targets, they can be the basis for mutually beneficial trade in entitlements.

Not all countries would benefit to the same extent from international trade in entitlements, and not all countries need to join in for large gains to accrue to participating countries. A regional climate change agreement could generate most of the gains for participating countries of global trade if it included countries tending to large imports of permits *and* countries tending to large export of permits. This could be built on the targets and rules for measurement, verification and reporting from the UN Framework Convention, but these do not need to be agreed in that multilateral context. If each member country were free to buy or sell emissions permits with countries outside the agreement, the regional agreement would be good for members while doing no harm to outsiders. There would be none of the damaging 'trade diversion' of traditional regional preferential trade agreements.

Assessing fair shares

So, in light of the new climate change regime, how do we assess each country's fair share of effort to reduce greenhouse gas emissions? We are seeking a practical answer to a practical question.

What is fair is in one sense what turns out to be acceptable for enough countries to make global mitigation work. Each sovereign state has to form its own judgment about whether and how much to contribute to a global mitigation effort. Explicitly or, more commonly, implicitly, it will go through the calculations that were undertaken for Australia in the 2008 Review and summarised in this book.

The world has groped its way towards the conception of what a fair share is in the Cancun Agreements. Developed countries have pledged to reduce the absolute amount of emissions by specified percentages. Major developing countries have pledged percentage reductions in the emissions intensity of production, or percentage reductions below business as usual.

These are not bad starting points. They need to be developed in a couple of ways.

While both developed and developing countries must enter commitments to constrain emissions, a distinction can be drawn on the way those constraints are set. An international agreement can work with developed countries accepting targets for absolute reductions in emissions, and developing countries targets for reductions in emissions intensity. What is missing are rules for the transition of a country from developing to developed status.

I suggested in the 2008 Review that the transition can take place when a developing country's emissions per person reach the (falling) average level of developed countries. That still seems practical as well as ethical.

The targets based on intensity and business as usual need to be defined precisely and quantified to make them verifiable and to provide a firm basis for international trade in entitlements. The intensity targets lend themselves more easily to precise quantification. They are a better basis for further development of the Cancun approach.

And a means has to be found for calibrating effort across countries: what percentage reduction in emissions in each developed country and in emissions intensity in a developing country represents a fair share?

The 2008 Review argued that a 'modified contraction and convergence' framework was the best approach to calibrating fair shares across countries.

This was based on the practical consideration that no basis for allocating entitlements to emissions would be broadly acceptable unless it allowed similar amounts of emissions to each person. It was not practical to move to equal entitlements per person overnight from the current position. The starting point on emissions per person ranges from Australia's 27 tonnes per person a year, to and below India's almost 2 tonnes (see Figure 3.1).

Figure 3.1: World Bank's chart of emissions per person in selected countries

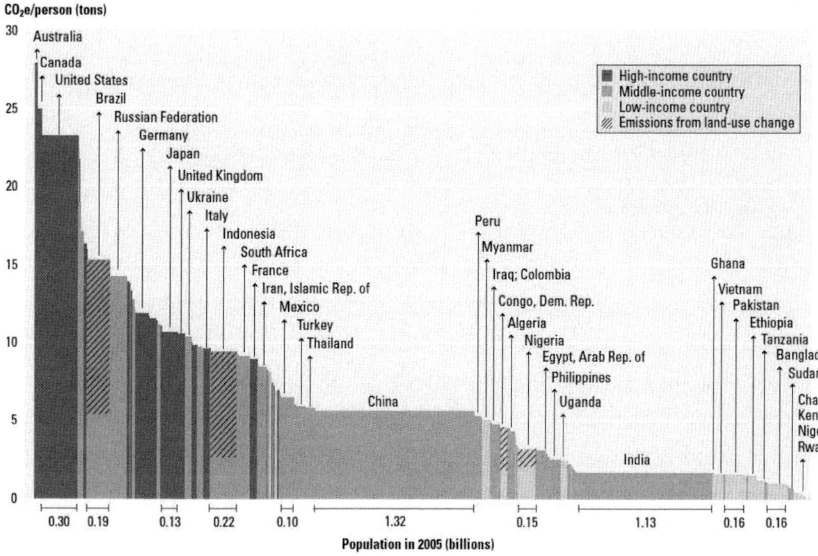

Note: The width of each column depicts population and the height depicts emissions per person, so the area represents total emissions. Emissions per person of Qatar (55.5 tonnes of carbon dioxide equivalent per person), United Arab Emirates (38.8) and Bahrain (25.4)—greater than the height of the y-axis—are not shown. Among the larger countries, Brazil, Indonesia, the Democratic Republic of Congo and Nigeria have low energy-related emissions but significant emissions from land-use change; therefore, the share of land-use change is indicated by the hatching.

Source: World Bank 2009, *World development report 2010*, p. 39.

Under contraction and convergence, the allocation for each country would move from its current level of emissions to equal entitlements per person in some later year. This is the convergence part. The contraction part is that total global emissions would fall from current levels to a much lower level at a specified date—a level low enough to meet the climate change objective.

It was not practical for developing countries, which were growing quickly with rapidly increasing emissions, suddenly to converge towards a low level of entitlements per person. The 'modification' part of 'modified contraction and convergence' is to allow rapidly growing developing countries to operate within an emissions intensity target for the time being. When they reached average emissions per person of the developed countries, their emissions per person would converge on the low emissions per person towards which other countries were moving.

The 2008 Review calculated a global emissions budget over time and allocated entitlements to draw on that budget within the modified contraction and convergence framework. Convergence would occur in 2050, by which time all countries would have entitlements per person below one-half of the current world average, or a bit below India's current level. This was broadly consistent with stabilisation (with overshooting) of emissions concentrations at 450 parts per million. This was the framework that gave rise to the 2008 Review's recommendation on targets for Australia and indicative targets for other countries.

The rationale for this approach is ethical as well as practical. It is ethical in that it does not place additional new hurdles in the way of raising living standards for millions of the world's poor at an early point in their countries' economic development. While efficient approaches to reducing emissions can hold costs to manageable levels, it is equitable that a higher proportion of those costs be borne by richer countries.

Two other considerations colour the ethical debate. There were two main criticisms of the 2008 Review's modified contraction and convergence approach, from commentators in India and China in particular. First, some commentators thought it unfair that those countries which happened now to have some of the highest emissions per person—Australia, Canada and the United States among developed countries—should continue to occupy that position for a long time into the future. Second, some critics thought that account should be taken of the historical reality, that developed countries had been responsible for most of the accumulation of greenhouse gases that has brought the world to its current dangerous position. There may be an emerging understanding that historical responsibility is handled best by developed countries assisting mitigation and adaptation in low-income developing countries.

It so happens that modified contraction and convergence gives similar results for Australia to percentage reductions from a base year. With contraction and convergence, Australia's high starting level of emissions causes the rate of reduction in emissions entitlements to be higher than it would be with simple percentage reductions from a base year. Australia's high population growth rate brings it down. Australians should recognise that the broad approaches to emerge from Cancun suit their national interests as well as any of the feasible alternatives.

The intensity approaches to emissions reductions pledged by major developing countries at Cancun are also broadly similar in structure to modified contraction and convergence.

Note that the targets for reducing emissions or emissions intensity relate to entitlements and not to emissions within a country's borders. This means that neither a country nor a firm operating within it is disadvantaged by having its exports concentrated in emissions-intensive industries—as Australia has with gas and coal and China with manufacturing—so long as two conditions are met. All substantial countries must accept targets; the way in which the targets are set is not important to the point under discussion. And there must be international trade in entitlements. The trade causes the carbon cost to be embodied in the world market price for the product, which allows the country and firm to recoup the cost of buying entitlements. Deep trade among a set of countries which includes major sellers and buyers of entitlements is enough to secure these benefits; not all countries need to participate in trade.

Lost alternatives

It has been suggested that quite different approaches to assessing fair shares would be better in principle or better for Australia than those that were the subject of the Cancun Agreements. The first of these is a 'consumption' rather than a 'production' approach to calculating emissions. The second is comparing explicit or implicit carbon prices rather than emissions levels.

Neither of these alternatives could be chosen unilaterally by Australia, and the established approaches are now deeply entrenched in international agreements to which Australia is a party. The alternatives are theoretical rather than practical possibilities.

One important reason cited in the 2008 Review for choosing a production over a consumption basis was greater ease of measurement and administration. This is still a consideration today.

Would a consumption approach be better for Australia if the world had gone that way? Certainly Australia, like China, has an unusually high level of emissions embodied in its exports. This affects the baseline from which changes in emissions are measured, as well as the changes in themselves. Over long periods, Australia's entitlements may be higher or lower with a consumption approach.

Would it have been better to use explicit or implicit carbon prices rather than emissions as a basis for comparing mitigation efforts?

Explicit prices would not serve. The international community decided long ago that countries would be free to choose their own preferred instruments for reducing emissions. Carbon prices have not been used as a major mitigation instrument in many countries, despite this being the means of reducing emissions at lowest cost. This reflects a range of political economy and administrative constraints, which would not be easily removed.

All interventions have an implicit carbon price—or, as the Australian Productivity Commission has pointed out, two implicit prices, relating to encouragement of supply of low-emissions production and discouragement of consumption. The careful work of the commission is demonstrating that implicit prices would be unsuitable as well as impractical as a basis for comparing different countries' efforts. The interventions are so numerous, and so varied across activities, that the calculations for each country are complex. If used as a basis for determining comparable effort across countries they would be contested. Once carbon prices, explicit or implicit, became a basis for international comparisons of effort, they could be easily manipulated by governments.

More fundamentally, it is difficult if not impossible to define what is a measure imposed to reduce emissions. In reality, governments seek multiple objectives in many policy measures that have the effect of reducing emissions. How much was the 'pink batts' subsidy introduced in Australia in the aftermath of the Great Crash a response to climate change, and how much to other things? When President Obama in his 2011 State of the Union address said that America should encourage clean energy to 'strengthen our security, protect our planet, and create countless new jobs for our people', did the multiple objectives dilute its value as a climate change measure?

We can be more specific. A proliferation of health and general environmental, as well as greenhouse gas, concerns have combined to make it virtually impossible for a construction permit to be granted for a new coal-fired power station in the United States today. In addition, some highly polluting coal-based generators are being forced into retirement. These are among the most important of the mechanisms through which the United States may reach its 2020 emissions reduction targets. If we are seeking to calculate the implicit carbon price in the United States, should these restrictions on coal-based power generation be included? The Productivity Commission had to answer this highly practical question in its study of carbon pricing. It chose to exclude these considerations. This is a defensible position. So would the alternative have been defensible.

Consider similar issues in China, relating to the much higher price of electricity to manufacturing plants that exceed some threshold in intensity of energy use and emissions. Or new instructions reported in the *People's Daily* in May 2011, that power and manufacturing industries exceeding specified emissions intensities would be closed. Again, the Productivity Commission's decision to exclude such measures from implicit pricing was defensible. At the same time, these are powerful mechanisms for reducing emissions. The decision could easily have gone the other way.

Outside the scope of the Productivity Commission's work, how would we view the withdrawal of the normal value-added tax rebates for the most emissions-intensive manufacturing industries, including aluminium, steel and cement? This was a climate change measure, but also met the various Chinese goals for reducing energy use.

It is impossible to draw distinctions on the basis of the motives of policy-makers.

What matters is not the motives, let alone the stated motives, for a policy decision that leads to reductions in emissions. What matters is the reduction in emissions. Facing up to this reality leads us back to comparing actual changes in emissions—absolute or intensity of production—as the best as well as the most practical way of comparing contributions to the global mitigation effort.

Some Australians argue that some of these countries—notably the two largest, China and the United States—have not adopted carbon pricing, and that Australia would be getting ahead of the world if it did so. Yet the mainly regulatory measures being taken by those countries impose greater costs on business and on their communities' standards of living than carbon pricing. This is clear from economic analysis. The Productivity Commission report on emissions pricing should provide empirical evidence on the costs of regulatory approaches to reducing emissions. While the higher costs of emissions reduction in other countries should not be counted as a contribution to the mitigation effort, neither should it count against them so long as they are meeting their commitments to constrain emissions.

Conclusion

Against all the odds, there is an international agreement on mitigating climate change. The world is on its way towards substantially reducing emissions growth.

There is a long way to go before the prospective costs of dangerous climate change have been reduced to acceptable levels. Success will come from building on current achievements of the international system, and not from starting again. Australia's strong national interest in effective mitigation is served by helping to make the emerging arrangements work.

4   Pledging the future

IN APRIL 2011, a small Australian power firm, CBD Energy, announced a $6 billion partnership with two large Chinese electricity firms, China Datang Renewable Power Co and Tianwei Baobian Electric Co. The joint venture plans to build a string of new wind and solar power plants across Australia.

The joint venture went ahead without any signed power purchase agreements with energy retailers and despite regulatory uncertainty surrounding carbon pricing and renewable energy targets.

The low cost of finance and technology from the Chinese side of the deal made the deal possible. Both China Datang and Tianwei Baobian are government-owned enterprises with ambitious goals for expansion into global renewable energy markets.

Australia's dedicated climate news service, Climate Spectator, described the deal as a game changer for Australian renewable power:

> If Australian companies can't get around to building their own renewable projects, then the Chinese will do it for them. Hu Guodong—vice president of Datang Renewable Power Co, the listed offshoot—says the slow rollout of renewable projects in Australia has presented an irresistible opportunity for companies such as his. 'Australia has amazing solar and wind resources.'

The deal captured the essence of shifting global leadership in climate change mitigation. While Australia has spent the last four years bickering, China has pledged large carbon intensity reduction targets, implemented reforms that deliver on its commitments, and set sail on a global mission to dominate new opportunities.

If we're to understand who precisely is doing what, let's begin with a look at their pledges.

The pledges

To date, 89 developed and developing countries, representing more than 80 per cent of global emissions and about 90 per cent of the global economy, have pledged large cuts and actions under the Cancun Agreements.

The quantitative pledges on 2020 emissions by a selection of major developed countries are listed in Table 4.1.

Table 4.1: Mitigation pledges to 2020 by selected major developed countries under the Cancun Agreements

Country or region	Pledge
Australia	• 5% reduction relative to 2000 unconditional • Up to 15% reduction if there is a global agreement that falls short of securing stabilisation of greenhouse gases at 450 ppm carbon dioxide equivalent and under which major developing economies commit to substantially restrain emissions and advanced economies take on commitments comparable to Australia's • 25% reduction if the world agrees to an ambitious global deal capable of stabilising levels of greenhouse gases in the atmosphere at 450 ppm carbon dioxide equivalent or lower
Canada	17% reduction relative to 2005; to be aligned with the final economy-wide emissions target of the United States in enacted legislation
European Union	20% reduction relative to 1990; 30% reduction as part of a global and comprehensive agreement, provided that: • other developed countries commit themselves to comparable emissions reductions • developing countries contribute adequately according to their responsibilities and respective capabilities
Japan	25% reduction relative to 1990, premised on the establishment of a fair and effective international framework in which all major economies participate and on agreement by those economies on ambitious targets
New Zealand	10% to 20% reduction relative to 1990, conditional on a comprehensive global agreement to limit the temperature increase to less than 2°C, with effective rules for land use, land-use change and forestry regulation, recourse to a broad and efficient international carbon market, and advanced and major emitting developing countries taking comparable action commensurate with their respective capabilities
Russia	15% to 25% reduction relative to 1990, conditional on appropriate accounting of the potential of Russia's forestry sector, and legally binding obligations by all major emitters
United States	Reduction in the range of 17% relative to 2005, in conformity with anticipated US energy and climate legislation, recognising that the final target will be reported to the UN Framework Convention Secretariat in light of enacted legislation

As is obvious from Table 4.1, in defining their pledges, countries have chosen different types of commitments and different base years. We can, however, interpret these commitments in terms of what would be a fair share from each country in the 2008 Review's modified contraction and convergence framework.

The pledged target ranges for the United States, the European Union and Japan all correspond to entitlements for a global agreement between 450 ppm and 550 ppm. The targets pledged by Canada and Russia, by contrast, are less ambitious than suggested for a 550 ppm global agreement. And, on average, developed countries' pledged 2020 targets are somewhat less ambitious than are needed under a 550 ppm scenario.

For developing countries, fair shares are measured not in absolute reductions but in reductions in emissions intensity (see Table 4.2). The modified contraction and convergence framework of 2008 implied a targeted reduction in China's emissions intensity of 35 per cent from 2005 to 2020 if global concentrations of carbon dioxide were to be limited to 450 ppm. At Copenhagen and Cancun, China pledged to reduce its carbon intensity by 40 to 45 per cent from 2005 to 2020. It thereby exceeded what was viewed as an adequate commitment even under an ambitious global agreement. India has pledged reductions in emissions intensity of 20 to 25 per cent on 2005 levels by 2020. Its proportional emissions intensity reduction for a 450 ppm outcome would have been 43 per cent. The Parikh report on low carbon growth strategies to the Indian Prime Minister in May 2011 commented that India could achieve emissions intensity reductions in the range 33 to 35 per cent with support from international financing and technological transfer.

The comparison of China's and India's 'fair shares' of a strong global agreement is determined by the arithmetic of modified contraction and convergence. The details of my 2008 formula are not the important thing, so long as the outcome is consistent with global goals. China's pledge exceeded its suggested emissions reduction by more than India fell short of its reduction. Experience has demonstrated that an alternative formulation of modified contraction and convergence is more realistic: China sets for itself an ambitious goal of reducing emissions intensity at a rate of 45 per cent over 15 years, and other rapidly developing countries go as close to that as possible.

A number of major developing countries have pledged reductions relative to a business-as-usual scenario (including Indonesia, Brazil, Mexico, South Africa and the Republic of Korea). Analyses of plausible business-as-usual scenarios have shown that, if realistic baselines are applied, the Copenhagen pledges imply reductions in absolute emissions in these countries between 2005 and 2020. These, too, are as ambitious—or more ambitious—than were called for under the modified contraction and convergence framework developed and proposed in the 2008 Review.

Overall, the modified contraction and convergence framework suggests that global commitments add up to somewhere near the level of reductions in emissions needed to limit greenhouse gas concentrations to 550 ppm. Within the same framework, developing countries are leading the effort in relation to their respective fair shares.

Table 4.2: Comparison of the Cancun pledges and notional entitlements under the 2008 Review's modified contraction and convergence framework

Country or region	Target type	Cancun pledges: change in absolute emissions at 2020 relative to 2000[a]	2008 Review: emissions entitlement allocations at 2020, relative to 2000–01	
			550 scenario	450 scenario
Australia		−5% to −25%	−10%	−25%
Canada		−13%	−33%	−45%
European Union	Reductions in absolute emissions	−12% to −23%	−14%	−30%
Japan		−33%	−27%	−41%
New Zealand		−27% to −35%	n.a.	n.a.
Russia		+15% to +31%	n.a.	n.a.
United States		−16%	−12%	−28%
Weighted average of developed countries		*−10% to −16%*	*−15%*	*−31%*
		Cancun pledges: reduction in emissions intensity, 2005 to 2020	2008 Review: reduction in emissions intensity 2005 to 2020, applying the Review's suggested approach[b]	
China[c]	Reductions in emissions intensity (ratio of emissions to GDP)	−40% to −45%	−35%	
India		−20% to −25%	−43%	

n.a. = not applicable.

a. Computations for developed countries (absolute targets): Countries' targets are converted from their chosen base years (see Table 4.1) to the 2000 base year used by the Review using estimates of total greenhouse gas emissions, excluding emissions from bunkers and land use, land-use change and forestry. The base year adjustment accounts for divergences from countries' submitted pledges as listed in Table 4.1, including the large divergence for Russia due to significant reductions in emissions between 1990 and 2000.

b. Computations for China and India (emissions intensity targets): The modified contraction and convergence approach articulated in the 2008 Review allows developing countries growth in emissions entitlements at half the rate of their GDP. In calculating emissions intensity to allow comparison with emissions intensity targets, this rule was applied for the period 2013 to 2020. The difference in required emissions intensity reductions between China and India is because of different rates of change in emissions intensity during the period 2005 to 2012 which are carried forward in an assessment under the Review's proposal. If the Review's rule of half the rate of GDP growth had applied from 2005, the 2005 to 2020 reductions in emissions intensity for China and India would have been 44 and 43 per cent respectively.

c. China's emissions intensity target only applies to carbon dioxide emissions.

The OECD's International Energy Agency formed a more pessimistic assessment, suggesting that existing commitments were heading towards 650 ppm. Frank Jotzo suggests that the Cancun commitments of others would trigger movement to minus 15 per cent in Australia within the conditional target entered at Cancun. The differences between the Garnaut Review and International Energy Agency assessments derive from different approaches to what happens after 2020. The point of difference is whether rapidly growing developing countries accept the suggestion that they should commit to reducing absolute emissions within a straightforward contraction and convergence framework once their emissions per person have reached the (falling) average emissions per person of the developed countries.

Countries making big pledges

While the US scientific community was instrumental in placing the global warming issue on the international policy agenda, Europe is at the forefront of policy action.

The Scandinavian countries were the world's first movers on substantial climate change mitigation policy and they have stayed in front. They have been pricing carbon since 1991. This is one reason why Norway is not a carbon-intensive economy, despite being the only other developed country with endowments of fossil fuels that are in any way comparable to Australia's. Norway's emissions per person are 10.9 tonnes, and Australia's 27.3 tonnes.

For those who fret about the effect of a carbon price on some generalised notion of Australian 'competitiveness' and who believe that measures of such things have meaning, Denmark, Finland, Norway and Sweden have all been higher than Australia on the World Economic Forum's Global Competitiveness Index every year over the past decade. In several years they have occupied three of the top four places among 139 countries, and in all years three of the top ten. Australia's ranking has ranged between 5th in 2001 and 16th in 2010, and has fallen as low as 19th. Australia has not ranked in the top ten since 2003. Norway, in particular, has played a leading role in providing support for mitigation in developing countries, including in Indonesia.

The European Union established an emissions trading scheme in 2005 and has steadily tightened its parameters since then. The European Union's conditional targets are relatively strong compared to those of other developed countries within the framework of modified contraction and convergence. Half the people in the developed world—half a billion people—are covered by the European Union's emissions trading scheme.

The major Western European economies, including Germany, France and the United Kingdom, have gone well beyond the mitigation requirements of the European Union of which they are members. The United Kingdom has recently confirmed a considerable increase in the ambition of its emissions reduction targets in the midst of continued economic pressures in the aftermath of the Great Crash of 2008.

The Conservative-led coalition government under Prime Minister David Cameron came to power with commitments to extend the strong mitigation policies of the Labour government that it replaced. The government's resolve was tested when the independent committee on climate change recommended the legislation of a target to reduce emissions by 60 per cent from 1990 levels by 2030. This is from levels that are already relatively low: in 2005, UK total emissions were 1.7 per cent of global emissions, compared to Australia's portion of almost 1.5 per cent. The committee's recommendations were debated at length in cabinet, and their acceptance was announced in May 2011. The new targets—50 per cent from 1990 levels by 2025—will now be binding under domestic law.

China took proposals for major reductions in emissions below business as usual to the Copenhagen meeting. It agreed to reduce the emissions intensity of output by 40 to 45 per cent from 2005 levels by 2020. China also committed to implementing the world's largest program of sequestration through forestry. Forest coverage was to increase by 40 million hectares and forest stock volume by 1.3 billion cubic metres from 2005 to 2020. And China pledged to increase the share of non-fossil fuels in primary energy consumption to around 15 per cent by 2020.

There have been suggestions after the event that China's carbon intensity commitments only reflect what China was doing anyway. That view cannot survive analysis of the economic realities or of the Chinese political economy. These goals were originally opposed by official advisers with responsibility for economic policy on the grounds that they may be unattainable, or attainable only at unacceptable cost to economic growth. Once they had been accepted by the leadership, it became the responsibility of the economic officials to make sure that they were achieved.

China is now implementing these commitments. The five-year plan for 2011–2015 approved by the National People's Congress in March 2011 announced a target of reducing the emissions intensity of national production by 17 per cent over the five-year period. The carbon intensity targets have been devolved down to provincial levels, and from there to local governments.

National officials have stepped in to override local government decisions that were thought to be inconsistent with the national objectives.

The authorities have pursued multiple environmental, energy security and other objectives by closing many emissions-intensive plants, and constraining energy supply or raising the cost of energy to others. There has been a rapid reduction in the emissions intensity of coal-fired electricity generation. Environmentally damaging, unsafe and economically inefficient small coal-fired generators have been closed at the rate of one every ten days or so. They have been replaced by larger plants that are economically and environmentally much more efficient. China decommissioned smaller, environmentally and economically inefficient plants with combined generation capacity of 70 gigawatts—one and a half times the total Australian power generation capacity of the electricity market.

Specific fiscal interventions have restricted or raised the costs of inputs to the most emissions-intensive industries, including steel, aluminium and cement. Some of these industries have been denied the rebates for exports that are normal in value-added-tax regimes in China, Australia and elsewhere, amounting to a large discriminatory tax on these emissions-intensive, trade-exposed industries. Some provinces now impose a surcharge on power use equivalent to either $19 per tonne of carbon dioxide on electricity used in highly emissions-intensive plants or $57 per tonne of carbon dioxide on electricity used in excessively emissions-intensive plants in eight 'high-polluting' industries. The high-polluting industries include aluminium, steel and cement.

There has also been substantial fiscal support and regulatory intervention to accelerate the deployment of a wide range of low-emissions technologies in energy and transport. The central government allocated $30 billion of its own funds to energy efficiency and emissions reductions projects in the five-year plan for 2006–10, generating total investment of $300 billion.

The interventions that have reduced growth in greenhouse gas emissions are varied and complex, but in total have had had a major effect on business costs and the way business is run. At the margins at which investment decisions are made, these have had large impacts. Over time, the interventions will reduce Chinese production and export capacity to well below where it would otherwise have been in aluminium, steel, cement and other highly emissions-intensive industries. This will tend to raise product prices on world and Australian markets.

The low-emissions technologies were a special focus of China's stimulus packages, adopted in late 2008 and early 2009 in response to the Great

Crash. There was massive support for deployment of virtually all of the low-emissions technologies: solar, wind, nuclear, biomass and hydro-electric. There was major investment in the electricity transmission grid to reduce energy losses and to facilitate integration of new sources of electricity. There was a major focus on accelerating the commercialisation of electric cars. There was huge commitment to expansion of public transport within urban areas, and extraordinarily rapid progress in developing 13,000 kilometres of fast train infrastructure to join up most of the large cities of China.

Not all of this went smoothly. There were examples of wind power capacity growing well in excess of the grid's ability to use the product. There was criticism by economists of wasteful levels of subsidy for deployment of rooftop solar and electric cars. But the overall effect was transformative. Problems have been identified and corrected. For example, the problems with absorbing rapid increases in wind power seem to have been reduced quickly by accelerated introduction of 'smart grid' technology. The Chinese grid authorities are following closely recent technological developments in the United States.

The Chinese economic policy authorities have been surprised by the rate at which the costs of the low-emissions technologies have fallen.

Costs of nuclear power have fallen so much that in coastal China nuclear is close to being economically competitive, with the relative costs continuing to move in its favour. The alternative involves the import of expensive coal from Australia and elsewhere or the expensive transportation of coal from the inland of China using hopelessly overextended rail and road systems. Soon the main constraint on expansion of nuclear at the expense of coal will not be cost, but anxiety about supplies of high-grade uranium oxide. Concern about nuclear safety issues in the wake of the March 2011 earthquake and tsunami in Japan has led to planned nuclear projects being suspended pending assessment of the recent Japanese experience. This is likely to lead to confirmation of the program with stronger safety standards.

Costs of wind power have fallen by one-fifth in two years despite the general inflationary environment in China. The cost of solar photovoltaic units has been decreasing rapidly and, as this is a younger technology, will continue to do so for some time.

Most Chinese mitigation so far has been through regulatory interventions. In August 2010, China's National Development and Reform Commission launched a national low-carbon province and low-carbon city experimental project. The eight cities and five provinces covered by the project will develop emissions reduction plans and explore options to use market mechanisms to achieve abatement goals. There has been discussion of linking emerging market-based arrangements to the European Union emissions trading scheme. China also plans to impose a new tax on coal, oil and gas extraction in its western provinces. The tax, introduced in June 2010 in Xinjiang, China's largest gas-producing province, will be broadened to include all western areas.

What once seemed unattainable targets to Chinese economic authorities are now viewed with confidence. Officials have been pleasantly surprised at the rate of decrease in costs and are now talking confidently of reaching the high point of the emissions intensity reduction. Figure 4.1 compares what China's emissions would have been under business as usual with the emissions projected under its Cancun target.

Figure 4.1: China's emissions under its Cancun target and business as usual (actual to 2010, then projections)

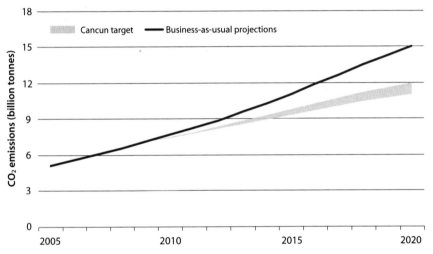

Source: Garnaut update paper 3, *Global emissions trends* (based on International Energy Agency 2010, *CO_2 emissions from fossil fuel combustion 2010*; IMF 2010, *World economic outlook—recovery, risk and rebalancing*; International Energy Agency 2010, *World energy outlook 2010*; BP 2010, *Statistical review of world energy 2010*).

China would not have committed itself to the targets offered under the Copenhagen Accord if they had been internationally legally binding, but senior officials are now suggesting privately that China may strengthen the nature of its commitments in the context of stronger international agreement.

China's actions represent by far the largest contribution to reducing global emissions below what they would have been under business as usual. These developments in China are the most important reason why uneven progress in most of the world since 2008 has not left a 2°C objective permanently out of sight. China's progress has also reduced the cost of capital items for deployment of solar and wind power in other countries. It may eventually do the same for nuclear power and rail transport.

This dimension of Chinese reality is confusing when placed alongside another: the fact that China's total emissions are still growing rapidly, and will continue to do so for some time.

The emissions numbers from China are daunting: they have become the largest in the world and continue to increase. But when we analyse them with clear heads they represent good news. China is doing what would be a fair share of a strong global effort, given its level of development and rate of growth.

China's rapid acceleration of growth in the 21st century will take its emissions per person to the (falling) average for developed countries in less than a decade. China's absolute emissions will then need to fall in line with those of the developed countries. The Chinese Government has not yet turned its mind to this next step.

Indonesia's pledges represent big steps. There is no doubting the commitments of leaders, but there are formidable political economy barriers to turning pledge into achievement. Indonesia has committed to reining in deforestation and improving land management in a bid to help fulfil its pledge to cut emissions by 26 per cent relative to business as usual by 2020. This pledge could rise to 41 per cent with international support. A moratorium on issuing new licences for land conversion was included as part of a US$1 billion agreement with Norway, and initiatives to improve institutions, incentives and monitoring in the forestry sector are under way. This includes the Indonesia–Australia Forest Carbon Partnership. Indonesia is also considering options for expansion of geothermal power production as a zero-emissions alternative to new coal-fired electricity generation. The Indonesian Government is considering financial incentives for investment in low-carbon power supply, and a carbon tax is currently being mooted in response to a 2009 Ministry of Finance climate policy strategy paper.

In 2008, Brazil committed through its National Climate Change Policy to reducing greenhouse-gas emissions by between 36.1 and 38.9 per cent by 2020, relative to business as usual. The policy pledges that fiscal and tax measures are to be introduced to bring about these emissions reductions. The Brazilian Government aims to reduce electricity consumption by 10 per cent by 2030 through a range of direct action measures. Much of Brazil's emissions reductions are likely to come through changes to land use. The Brazilian Government aims to reduce deforestation to 80 per cent of the annual average between 1996 and 2005 and to double the area of forest plantation by 2020.

Experience and experiments with carbon pricing

As with the Kyoto Protocol, countries that pledged targets or actions under the Copenhagen Accord were free to determine the policy measures they would put in place to achieve their pledged targets or actions.

More than half of the population of the developed world lives in countries with emissions trading schemes. More than 30 countries have introduced, or are seriously considering introducing, market-based measures to help meet their emissions reduction targets affordably and efficiently. Several countries' carbon-pricing mechanisms include design features that allow the rate of emissions reductions to be accelerated if other countries take on more ambitious targets. Countries and regions that have implemented carbon-pricing mechanisms—for example, New Zealand and the European Union—are considering various options to link up their trading schemes.

The European Union emissions trading scheme operated under an explicit trial phase between 2005 and 2007 and is currently in its first full phase, which runs from 2008 to 2012. The scheme covers more than 11,500 installations, which represent around half of Europe's carbon dioxide emissions and about 40 per cent of its greenhouse gas emissions. The 27 European Union member states, plus Iceland, Liechtenstein and Norway, are covered by the scheme.

New Zealand's emissions trading scheme started in 2008, initially covering forestry. Transport fuels, electricity production and industrial processes were added on 1 July 2010. Transitional measures in place between July 2010 and December 2012 allow participants to buy emissions units from the New Zealand Government for a fixed price of NZ$25. At the same time, participants in the energy, industrial and liquid fossil fuel sectors are required to surrender only one emissions unit for every two tonnes of emissions they produce. The combined effect of these measures is to cap the price of emissions units at NZ$12.50 until the end of 2012.

In 2010, the Japanese Government announced its intention to pass legislation that supported an emissions trading scheme, a carbon tax and feed-in tariff measures. However, in late 2010, the government announced that it would delay, but not scrap, plans to implement an emissions trading scheme. A number of voluntary schemes exist in Japan. These include the Japanese voluntary scheme, which was launched in 2005, and the experimental scheme, launched in 2008. The Tokyo Metropolitan Government also launched an emissions trading scheme in April 2010. The Japanese private sector has taken big steps to put itself in a leading position technologically in the development of products and processes that are suitable to a low-carbon economy.

In December 2009, the Republic of Korea's National Assembly passed the Framework Act on Low-carbon Green Growth, which paves the way for a mandatory emissions trading scheme. While the scheme is still being developed, the Korean Presidential Committee on Green Growth has indicated that the first phase of the scheme may run from 2013 to 2015.

A number of European countries beyond the Scandinavians, including the Netherlands and Switzerland, have implemented carbon taxes.

The South African Government is also considering how to introduce a carbon price. Its National Treasury released a discussion paper in December 2010 exploring the economic rationale for, and possible approaches to, introducing a tax on carbon.

In July 2010, India imposed a clean energy tax of 50 rupees per tonne (approximately A$1.13 per tonne) on both imported and domestically produced coal. Revenue from the tax will fund research and projects in clean energy technologies.

The three high-emissions developed countries

Australia, Canada and the United States have the highest emissions per person of the developed world. The economic structure and pattern of political interests associated with exceptionally high emissions have made it difficult for these countries to break away from old patterns of energy use. The result has been that they have held back the global mitigation effort.

There has been a tendency for people in these three countries who want to avoid action to look to the other laggards for comfort. The United States is obviously more influential on this issue than the other two countries, although Australians probably underestimate the extent to which their discussions and decisions play into the American debate. Over the past year, US officials close to the president have emphasised to me the significance of Australian progress in pricing carbon to the prospects for the use of economically efficient approaches to mitigation in the United States.

Canada has now hitched its mitigation ambitions to the American wagon. Its Copenhagen commitment is to reduce 2005 emissions by 17 per cent by 2020, unless the United States' objective is varied. While the US and Canadian mitigation goals fall well below those of many other countries, and while they fall well short of the requirements of current international climate objectives, they do represent a marked departure from historical trends.

The 2008 Review demonstrated that Australia has a more acute and urgent interest in the success of climate change mitigation than the United States and Canada—indeed, than any other developed country. There is therefore an awful incongruity in Australia taking comfort from the Canadian position in particular. Australia has a stronger interest than the others in trying to encourage all three of the high emitters to make positive contributions to the global effort.

As a close friend and ally of the United States, Australia has good reasons to look beyond narrow and specific reciprocity on climate change policy. There are many areas of common interest in which the United States carries disproportionate costs. This is true of much of the two countries' shared security interests. If it happened that in one area of shared interests, climate policy, the United States found it difficult to fully reciprocate an Australian contribution, there would be good reason for Australians to understand this as part of the fabric of a larger relationship from which it derives big benefits.

However, despite the considerable current domestic political difficulties on climate change mitigation policies, the United States is far from standing still.

A shift to a global carbon market suffered a blow when the Obama administration announced it would not pursue the passage of federal cap-and-trade legislation in 2010. But there have been considerable regional developments, with ten north-eastern and mid-Atlantic states now participating in a regional emissions trading scheme—the Regional Greenhouse Gas Initiative.

On 16 December 2010, the California Air Resources Board approved a cap-and-trade plan for California to be implemented in January 2012. Only ten national economies are larger than California's. This emissions trading scheme will be the world's second largest (after the European Union's) and aims to cut California's emissions to 1990 levels by 2020. This roughly corresponds to a reduction of 6 per cent from 2000 levels.

Of greater immediate significance, the US Environmental Protection Agency is pursuing aggressive regulatory measures, such as tightening regulatory restrictions on emissions from vehicles and mandating the closure of the most heavily polluting coal-fired power stations. Strong support for low-emissions sources of energy has been a feature of budget programs since the early stimulus packages in response to the Great Crash of 2008.

In all of these measures, the Obama administration is following the reputable scientific community. With a Nobel laureate in physics, Stephen Chu, as energy secretary, the cabinet is confidently plugged into mainstream scientific thought. It is working on the basis that climate change is a reality, that human activity is contributing influentially to it, that the human community faces large risks of disruption to its economic and political life, and that the problem is an urgent one.

The administration has remained committed to strong outcomes from international climate change negotiations. President Obama played a central role in the development of the Copenhagen Accord, and the United States was crucial in its consolidation and extension in the Cancun Agreements. The United States has left its 'minus 17 per cent' emissions reduction target on the table as a commitment under the Copenhagen Accord, necessarily qualified by references to US domestic processes. It has indicated that it will meet its commitments to the funding mechanisms established at Copenhagen and Cancun.

US officials at the highest levels state that the emissions reduction target will be met, despite the absence of a national market-based instrument for securing that result. They are supported in their statements of confidence by a number of factors. The slower economic growth that has followed the Great Crash helps a bit. The significant state-based initiatives add up to something that is noticed. The United States is enjoying a 'gas revolution', through which the competitive position of lower-emissions gas has been greatly strengthened against coal by a historically exceptional expansion in gas reserves. Finally, regulatory interventions at the federal level are becoming much more important. Let us now look at the last two of these.

The expansion of gas reserves has taken the United States by surprise. In the United States, the increase in gas reserves is associated most strongly with technological developments that have reduced the cost of large-scale gas recovery from shale deposits. A recent major study of the US gas position by the Massachusetts Institute of Technology estimates that gas reserves after depletion increased by 77 per cent from 1990 to 2010. This expansion of reserves has already reduced both average prices and their volatility, making gas a much more competitive fuel for power generation.

The opportunity for rapid expansion of the use of gas for power generation has been enhanced by the rapid expansion of gas generation capacity over recent decades. Since the removal in 1987 of various regulatory restrictions designed to preserve gas for what were thought to be socially more valuable uses, the United States added 361 gigawatts of power generation capacity, of which 70 per cent was gas-fired. Much of the new gas capacity was underutilised through a period in which gas prices were tending upwards with oil prices. It is now relatively easy and cheap to switch from coal-fired to gas-fired power generation. With greater regulatory pressure to close 'dirty coal' units, including some that are sources of high concentrations of mercury and particulates as well as greenhouse gases, there is considerable value in switching to gas-based power generation.

In recognition that there will be no market-based system of emissions reduction for the time being, the administration has increased the profile of regulation. Federal regulatory interventions are lifting the priority of emissions reduction and more generally the shift to clean energy. Stimulus spending in the aftermath of the Great Crash included programs for clean energy research, renewable energy deployment, public transportation, vehicle electrification and smart grid technology totalling US$67 billion. These themes have been continued in subsequent administration policy. The centrepiece of new policy in President Obama's State of the Union address in 2011 was a commitment to raise the proportion of 'clean energy' in US electricity generation from 40 per cent to 80 per cent by 2035.

The path for a wider role for regulation through the Environmental Protection Agency was cleared by a Supreme Court decision in 2007 that greenhouse gases fit within the United States Clean Air Act definition of 'air pollutants'. Coupled with the agency's finding in 2009 that greenhouse gases threaten public health and the welfare of Americans, the way was paved for promulgation of greenhouse gas emissions standards for new cars and light-duty (passenger) vehicles. The new standards will apply from vehicle model year 2012, and are estimated to reduce emissions from the United States light-duty fleet by 21 per cent by 2030, relative to what they would have been in the absence of the regulations (business as usual). In May 2010, President Obama issued a memorandum to expand the scope of the regulations to cover medium- and heavy-duty vehicles, starting with the 2014 model year.

Again in recognition that in the absence of a national carbon pricing mechanism such regulatory measures will be the central instruments for reducing emissions for some time, and that they are costly, the administration is seeking to reduce costs by introducing consistent approaches to evaluating policies. A 'social cost of carbon' has been developed that is applied in assessing proposals for regulations. The resulting number, or rather numbers, as there is a range thought to be appropriate across various circumstances, are now being systematically applied in decisions on the regulation of emissions from vehicles, appliances, and power generation and industrial facilities. The 'central' price was US$21 per tonne of carbon dioxide in early 2011. This price will be reviewed regularly.

The regulatory powers of the federal government are reducing greenhouse gas emissions and are likely to contribute much more in future. They are being challenged by the Republican majority in the House of Representatives, so far without effect.

The Obama administration included in its support for clean energy the provision of loans for new nuclear power plants. No new nuclear power plant has commenced construction in the United States since 1977. Despite fiscal and political encouragement under the Obama administration, progress remains slow. There is strong opposition from communities that have not lived with neighbouring nuclear plants, but there is often more support for the expansion of established plants. Anxieties have increased as a result of the developments in the Japanese nuclear industry following the March 2011 earthquake and tsunami. The economics of nuclear power generation have been set back by the new abundance and low cost of gas.

In an environment of regulatory focus on non-greenhouse gas pollutants from coal, local political activism against coal-based power generation, uncertainty about future pricing of emissions and low gas prices, investment in new coal-based power generation has become unlikely. Independent organisations have assessed that established regulatory measures and other policies could, in favourable circumstances, reduce US emissions by up to 14 per cent below 2005 levels by 2020. These studies do not take account of the gas revolution. The wide range of developments described above make it possible that the United States will achieve its 2020 emissions reductions targets despite the absence of economy-wide pricing of emissions. Of course, much will depend on the evolution of the national political balance in the years ahead.

Australia's fair share

The Australian Government and Opposition accepted the 2008 Review's proposal that Australia should reduce emissions by 5 per cent in 2020 from 2000 levels whatever the rest of the world was doing as our contribution to keeping hopes for a strong international agreement alive. The Review recommended—and the government accepted—that Australia should also make pledges of stronger commitments calibrated to what other countries were doing. Australia should offer to reduce 2020 emissions by 25 per cent in the context of a strong international agreement focused on holding concentrations at 450 ppm corresponding to a temperature increase of about 2°C.

If the world had reached effective agreement on emissions reductions that would lead to concentrations of 550 ppm, our fair share would have been 10 per cent.

The conditional targets are crucial.

The worry is that, on current trajectories, Australia would overshoot its Cancun 2020 target (even the minus 5 per cent target) by much more than other countries. That would be damaging to the global mitigation effort.

How soon should we move our unconditional target of minus 5 per cent, and how far, in the light of others' pledges and actions?

Given our starting point, the realistic ambition is to catch up with our fair share, rather than to be a leader.

The range for conditional targets recommended in the 2008 Review and accepted by the Australian Government still seems appropriate.

The time to adjust targets would be following the government's receipt in 2014 of the first report of the proposed independent committee (see Chapter 5). It would be appropriate for the target to be the percentage reduction of emissions which, within a contraction and convergence framework leading to equal entitlements per person in 2050, represents the average percentage reductions of the developed countries. The average would be weighted by population. The percentage would be based on pledges but adjusted for evidence of underperformance or overperformance against targets. The calculations would presume (as is currently the case) that the pledges of major developing countries add up to an equivalent reduction of emissions within a modified contraction and convergence framework. If they do not, there can be an appropriate adjustment of the target.

Conclusion

Perhaps the most extraordinary feature of the Australian public discussion of carbon pricing in 2011 is the common assertion that if Australia were to do anything to reduce emissions it would be acting alone. The assertion is often accompanied by statements that it would be economically damaging for Australia to act ahead of the world.

When you next hear someone say that he is worried that Australia might get ahead of the rest of the world in reducing greenhouse gases, take him by the hand and reassure him that he has no reason for fear.

There is no risk of Australia becoming a leader in reducing greenhouse gas emissions—others are already too far ahead. But we do run the risk of continuing to be a drag on the global mitigation effort. That is not a clever position for Australia—the developed country that is most vulnerable to climate change, and which is going through a once-in-history boom in incomes.

It would be a reasonable aim to be making good progress in catching up with the average of the developed countries. And we do have a chance of getting ahead of the pack in the way we go about reducing emissions. With carbon pricing we can do as much as others at lower cost. That is one way of getting ahead of the world that shouldn't frighten anyone.

PART II
AUSTRALIA'S PATH

5 Correcting the great failure

AFTER THE long, slow slide in the national wealth rankings that characterised much of Australia's 20th century economy, our century of protectionism ended in a series of measures undertaken from 1983.

The move coincided with a similar shift in China. And it was followed within a decade by several Southeast Asian economies, and more ambivalently (but in the end decisively) by India and by the collapse of the Soviet Union. Within a stunningly brief historical period, the world had its first truly global economy since the 19th century.

Australians did not want to join this global movement. No Australian industry wanted to be pushed from the safety of local conditions into the tough world of great and unknown competitors. But once the barriers were down, a remarkable reinvigoration happened. Australian productivity surged.

All exports grew strongly, but were led by an extraordinary range of goods and services, many embodying high intelligence and skills. Manufacturers, services and raw materials producers all flourished in unexpected ways. Education moved from being an inward-looking home industry to the top ranks of our export industries. Australian firms established dominant global positions in niches as diverse as insurance (QBE), infrastructure finance (Macquarie), blood products (CSL), cardboard boxes (Amcor), shopping malls (Westfield), travel publishing (Lonely Planet) and surf clothing (Billabong).

The wonders of the free market, and the inspiration, energy and hard work of the Australian private sector, took businesses to global success that neither politician nor bureaucrat could have picked.

It was and remains a remarkable period of Australian corporate reinvention built upon the innovative genius that free individuals bring to their craft. Such are the forces that can be unleashed when individuals are given the appropriate signals.

A similar historic choice confronts Australia now in its goal to reduce greenhouse gas emissions. On the one hand, a market-based price on emissions reflects the costs that atmospheric carbon imposes on the rest of society and asks individuals and firms to adapt and create solutions that incorporate that price. The other approach is by regulation, through which firms and individuals are required by law to refrain from emissions-intensive activity to an extent that adds up to the required reduction in emissions. In the latter approach, the government controls many consumption and production decisions by individuals and firms, based ideally on careful calculations of the activities that can reduce emissions at the least social cost.

The market-based approach requires all of the information that determined the recommendations on the targets for reducing emissions that underpin this book. The regulatory approach requires all of the information required by the market-based approach. It also requires a lot of information about individuals' and firms' responses to intervention by government and about the costs of those interventions.

Australia was only one small part of the 20th century struggle of ideas about ways of managing the economy. The regulatory approach went under the names of 'central planning' and 'protectionism'. The case for regulation depended on assessments of high transaction costs and instability in the market economy, as well as on the capacity of government to make a wide range of decisions more reliably than individual economic actors.

As it was here in Australia, that contest of ideas was won everywhere by the market economy. It was not won in theory. It was won by observing the results of predominantly market-based decisions and predominantly regulatory interventions. The market economy proved itself able to create solutions and find opportunities with far greater efficiency than the regulatory approach.

It is not that the outcome of the contest disqualifies interventions of some kinds, where that is clearly the most effective way of correcting specific market failures. But it has left a presumption in favour of market-based decisions unless there is clear evidence that regulation would give better results in a particular case.

As noted by Nicholas Stern, climate change represents the greatest market failure the world has ever seen. With a price on carbon, individuals and businesses can take into account the costs of their actions that are borne by society at large. Individuals and firms can decide how emissions will be reduced to meet Australia's fair share in global emissions reductions. Millions of people will find millions of ways—large and small—of reducing emissions at relatively low cost. They will find ways that no politician or bureaucrat in Canberra, Washington or Beijing has ever thought about. They will use specific and often local knowledge to discard some that might quickly have crossed the minds of the bureaucrat. The introduction of a carbon price to correct for the external costs of emissions in itself is an economic reform where the benefits far outweigh the costs.

Putting a price on carbon is not the whole climate change mitigation policy story. There are some other market failures pertaining to incentives associated with the carbon price. The most important of these makes a case for the provision of public support for investment in research and development of low-emissions technologies.

The carbon price operating through markets leads to changes in decisions that used to take no account of the costs of climate change. But it is actually a less distorting and less economically costly form of taxation than many of the other ways in which Australian governments raise revenue. A judicious use of the revenue raised by pricing carbon can increase economic wellbeing to the extent that it is used to reduce other highly distorting taxes.

The carbon price

Currently, the global emissions reduction challenge is to implement the world's commitment to limiting temperature increases to 2°C (or 450 parts per million carbon dioxide equivalent).

The 2008 Review recommended that, in the absence of an effective global agreement, Australia should introduce an emissions trading scheme geared to what others were doing but begin with a fixed price period. We would then be ready to define a target and to float the emissions permit price at some later time, when there were clear rules and opportunities for international trade in permits.

Several clear principles must guide policy if a carbon price is effectively and efficiently to drive the transition to a low-carbon economy. The principles are derived primarily from the objective of the policy—that is, to address the market failure of the cost that one firm's greenhouse gas emissions imposes on others.

- environmental integrity—confidence that genuine emissions reductions have been achieved on the scale required
- cost-effectiveness—emissions reductions should be achieved at least cost to the community, by avoiding duplication and overlap with other policies, and using revenue from the scheme to reduce the costs of mitigation
- swift revision of the scheme in response to the recommendations of regular, transparent and independent reviews—sound, independent governance will increase the chances that the scheme moves to its optimal design over time
- autonomy—the scheme should minimise reliance on recurring judgments by government, and instead harness the efficiency of the market within an independently managed framework.

Other important criteria for assessing options for carbon pricing models include administration and transaction costs; the ability to provide confidence for investors and participants; and the opportunities to support, and link to, existing and emerging international markets.

There are several models for putting a price on carbon. All are reasonably described as market-oriented approaches. The major difference is that some models set limits on the quantity of emissions and allow the price to vary, while others set the price of emissions and allow the quantities to vary. That said, carbon-pricing models share core features, including their use of a price signal and promoting greater efficiency benefits than regulation. Most, but not all, generate government revenue.

An emissions trading scheme with an initially fixed (and rising) price has some advantages. In the short term, a fixed price can provide steadiness, when a floating price would be volatile while the scheme remained the subject of fierce political dispute. It allows firms to become familiar with compliance under the scheme, and allows Australia to move towards a quantity constraint as knowledge of the scheme and confidence in its stability expand. Such an approach has the added benefit of gradually building industry capacity, and establishing and testing the necessary institutions and administrative infrastructure.

International trade in abatement is a legitimate and important element of an efficient global solution to climate change. The eventual transition from a fixed to a floating price, as well as linking with other schemes, will assist in allowing emissions reductions to take place where they are cheapest. One advantage of emissions trading over a carbon tax or an emissions trading scheme with a permanent fixed price is that it facilitates just such private international trade. Otherwise trade in entitlements has to be conducted through a government window.

Australia's resource endowment and comparative advantage in emissions-intensive industries makes our country a natural importer of permits and exporter of emissions-intensive products. We have fewer opportunities for low-cost abatement on the scale required to meet reasonable targets than many other countries (although development of land-based opportunities may change this, as discussed in Chapter 10). With greater opportunities for trade in emissions entitlements, Australia can be more ambitious and commit to doing its fair share in global action at lower cost.

In implementing an emissions trading scheme with a fixed-price start, there are two sets of decisions to be made: the starting price and how much the price will rise in each subsequent year; and the timing, conditions and manner of transition to emissions trading with a price that is set by market exchange.

The first objective of Australian mitigation policy must be to support the emergence of a strong and effective global agreement. This must be kept in mind in setting a domestic carbon price. The price must be consistent with Australia contributing its fair share to the global effort to reduce emissions. It should set us on a path to meeting the commitments to reduce emissions that we have made to the international community.

The setting of the initial price should also put Australia on a path towards longer-term outcomes. We need to place our economy in a good position for the future emissions reduction challenge and a world of global action. We should ensure that we do not encourage arbitrary or redundant investments or divestments that make no sense in the carbon policy world that follows the fixed price.

If Australia's carbon price is set too high—out of step with international action—there could be an unnecessarily costly transition. This is likely to raise doubts about the scheme's sustainability. Expectations that the scheme may be amended or abandoned will raise the supply price of investment in activities affected by it.

On the other hand, too low a price could impose transactions costs for no real gain. It would not raise the chances of reaching the goals of Australia and the international community. In the absence of a logical link to the larger objective, it would be difficult to establish credibility.

Australia's current policy settings and commitments are also relevant to a starting price. Australia has had an unconditional target since 2008 to reduce emissions by at least 5 per cent by 2020 (relative to 2000 levels). This target has bipartisan support. It became a commitment to the international community in Copenhagen in December 2009, and became part of a set of international agreements at Cancun in December 2010. Modelling suggests that to meet this target Australia's carbon price would need to commence at around $26 in 2012.

The targets allow unlimited permit imports, so Australia's domestic emissions could exceed the number suggested by its target. The use of imported entitlements must depend on the integrity of the available international permits—they must represent real reductions in emissions in partner countries with hard targets and must not be counted against the targets of the countries from which they have been purchased. Trading partners should have a firm national target, whether calculated on a percentage reduction of emissions in a base year or a reduction in emissions intensity. The 5 per cent figure is a 'net' rather than domestic emissions reduction.

We will need to tighten our target in line with international action, hopefully to the levels that would be required if the international community is to reach its declared goal of holding the global temperature increase to 2°C. The starting point has to prepare us for later adjustment if and when it is required.

Also relevant are explicit carbon prices in existing international markets and places where economy-wide carbon pricing policies are present. Future linking and trade in entitlements will occur more smoothly if the gap between Australian and overseas carbon prices is not too great. The current (May 2011) price of emissions permits in the European Union emissions trading scheme is around €17 (A$23) per tonne of carbon dioxide equivalent. The current price of offsets in the form of Clean Development Mechanism credits is around €13 (A$17) per tonne.

Another indication of suitable, and credible, prices for carbon is provided in economic analyses that guide regulatory decisions in the United States, where a systematic approach has been taken to these issues. The US Government recommends that economic assessments use a social cost of carbon of US$21 (A$20) per tonne of carbon dioxide equivalent, rising over time to US$26 (A$25) in 2020, and US$33 in 2030 in 2007 dollars (A$31). In the United Kingdom, this price is higher, with investors in the non-traded sector advised to consider £26 (A$40) per tonne to be a suitable cost of carbon.

Taking all of these considerations into account, I recommend that Australia's initial carbon price be in the range of $20 to $30. The mid-point of this range would be appropriate in the absence of compelling reasons to move away from it.

Once a carbon price is established, its rate of increase will need to balance the considerations outlined above: Australia's contribution to global goals, our existing commitments, domestic credibility and other countries' climate change mitigation policies and their associated implicit carbon prices.

Prices to ensure the optimal depletion of a finite resource—in this case, the earth's limited absorptive capacity—will increase over time at the rate of interest, as Hotelling concluded back in 1931. It is my assessment that a mature market would come to apply something like an interest rate of about 4 per cent in real terms—2 per cent representing the risk-free real rate, and the other 2 per cent a risk premium. This is the rate at which a well-informed market could be expected to raise the rate over time if the initial rate had been set appropriately to meet an emissions reduction target that is not changed over time. It is appropriate, then, to simulate the likely market movement by raising the fixed price of emissions by 4 per cent per year in real terms.

Floating the price and setting the target

Investors need clarity about when and the conditions under which the transition to a floating price will occur. To support a smooth transition, the necessary institutions and supporting infrastructure should be established from the beginning of the scheme. It is important to specify rules for the scheme as soon as possible, including arrangements for auctioning permits and for acceptance of offsets and international permits. Having this framework agreed, understood and embedded alongside a fixed price will build confidence in the transition, and allow rapid and smooth movement to a floating price when the time is right.

The following conditions might be considered to be relevant to the timing of a shift to a floating price:

- Development of global agreements if sufficient countries (weighted by significance in the international economy and trade) take on emissions targets in the medium and long term. Following Cancun, this condition would seem to have been met, although it may be wise to wait and observe for a while the implementation of the Cancun agreements.

- Opportunities for trade. These may exist in substantial quantities, liquidity and stability in advance of the kind of global agreement envisaged in the 2008 Review. They could be nurtured through a regional agreement with neighbouring countries that are complementary to Australia. A regional market would need to be underpinned by emissions targets that represent each member's fair share in a global effort, and in the short term by commitments that are proportional to comparable countries' commitments. Trade with New Zealand, Indonesia, other ASEAN countries, Japan, Korea and Indonesia, and potentially Papua New Guinea, Timor-Leste and the South Pacific, may be relevant. Australia and Indonesia could discuss the merits of forming the core of a wider regional agreement. In developing countries with weak administrative systems, other countries would need to provide assistance with administration and compliance. Trade with the European Union and parts of North America may become relevant. However, the latter would probably be similar to Australia in seeking to purchase surplus emissions entitlements from others, and so trade with them would need to be within wider trading arrangements that included countries that were naturally net exporters of permits.

- The establishment of credibility and stability of the domestic scheme. The desire to build confidence during a period of political uncertainty is one reason for starting with a fixed price, and this role is completed when the domestic political process has accepted that the scheme is here to stay.

Judgments about whether the above conditions have been met will have subjective elements. It will be difficult for participants in the market to assess when the transition might occur. There is a risk that uncertainty would encourage destabilising pressure on the decision-making process by interests that stood to gain or to lose from a delay in the transition to a floating price.

On balance, therefore, there are advantages in fixing the date of transition in advance, and in working to ensure that adequate opportunities for credible international trade in entitlements are available by the time of transition. This book favours three years, that is, in the middle of 2015, unless the independent regulator, on expert advice, judges that the opportunities for international trade in entitlements are not sufficient to support a liquid and stable permit market.

A firm target for reductions of emissions over time will need to be established in advance of the movement to a floating permit price. Australia's current unconditional target for 2020 would be the legislated minimum emissions reduction target. The setting of a target above the minimum should be considered in the first two years of the scheme, following the first independent review of the target. Independent reviews should occur regularly, on a pre-announced set timetable.

The process and institutional arrangements for such a review are important. In the United Kingdom, the Climate Change Act 2008 mandates an emissions reduction target for 2050, and the processes for setting interim carbon budgets. The act requires the government (through its secretary of state) to take into account the advice of the independent Committee on Climate Change (established under the act), along with any representations made by other national authorities. The committee's input includes advice on whether the 2050 target should be amended, and on the level for interim carbon budgets. The legislation requires that if the government sets the carbon budget at a different level from that recommended by the committee, the secretary of state must also publish a statement setting out the reasons for that decision.

Governing Australia's emissions trading scheme

There will be no success in mitigation, at a national or international level, without good governance. The policies that will mitigate climate change cut across strong interests of many kinds. These are circumstances in which it is easy, indeed natural, for vested interests to capture policy, and for the ultimate reasons for policy to be forgotten. Good governance is an antidote to these tendencies: the articulation of clear and soundly based principles as a foundation for policy, and the establishment of strong, effective and well-resourced institutions to implement the principles.

I have recommended that three independent bodies be established to implement and administer Australia's carbon price arrangements: an independent scheme regulator, an independent committee to advise on targets, and an independent agency to advise on trade-exposed industries.

The scheme regulator, or carbon bank, should have a high degree of independence in the exercise of its responsibilities. The overarching objective of the carbon bank would be the implementation of the scheme as established in legislation. The carbon bank would also administer the assistance to trade-exposed industries.

An independent committee, similar to the UK Committee on Climate Change, would provide advice to the government on national targets and scheme caps; progress towards meeting targets; the switch to a floating price; and expanding coverage of the scheme. This would be done through regular reviews of the scheme, with the first review to occur no more than two years after commencement of the scheme so that its advice is available to the government before the switch to a floating price for emissions, and with subsequent reviews no later than five years after the preceding review.

As with the setting of emissions targets in the United Kingdom, the government could retain power to override operational decisions of the independent authority, provided that a statement is made to parliament within three months outlining and explaining its decision. The head of the authority would periodically appear before a parliamentary committee. In any case of adjustment, a new target would be announced promptly after the receipt of the independent advice, and legal adjustments made with effect no later than two years after the scheduled date of the review. The first review should be completed and the initial advice provided to government within two years of the commencement of the scheme. Reviews should occur at intervals no greater than five years.

There is good reason to expect sufficient trade to be present to switch from a fixed to floating price in 2015, so long as the establishment of international trading arrangements is given high priority by the government. In the remote circumstance that this does not come to pass, the independent body should examine the case for continuing the fixed price arrangements, taking into account this issue along with other relevant factors.

Voluntary action

Voluntary emissions reductions by households and businesses should receive recognition in the administration of compliance with targets. As proposed by the government in its final Carbon Pollution Reduction Scheme package in 2009, this can be achieved by allowing for voluntary purchase of offsets for emissions (for example, for air travel), 'green power' and other similar arrangements to be added back into the emissions base for purposes of compliance with international commitments. In this way, the voluntary activity leads to a commensurate increase in the ambition of the emissions reduction target.

Conclusion

This is the fourth time that Australia has moved towards economy-wide carbon pricing. Each time, the retreat of economy-wide action did not mean the end of climate change mitigation policies. An array of regulatory interventions took their place, with little effect on emissions but large effects on the Australian standard of living.

The US Government is, for the time being, adopting a relatively expensive approach to reducing emissions because it has no choice. Some Australians make that an argument for Australia to follow the United States in adopting relatively expensive means of reducing its emissions.

American economist Jagdish Bhagwati used to characterise a similar common argument for trade protection as: 'Beware. I will keep shooting myself in the foot until you stop shooting your own feet'.

If we are clever, we can apply mitigation policies that have relatively little effect on the rise in living standards in the years immediately ahead. We can do so while contributing our fair share to international action that provides substantial protection for the Australian standard of living in the more distant future.

The alternative is to suffer a major setback to productivity and the rise in living standards—now, from expensive mitigation policies; or later, as we face the consequences of failure of the international mitigation effort.

Australians would do well to make sure that this fourth movement towards a carbon price corrects Australia's part of the great market failure.

6 Better climate, better tax

I WAS FLYING back across the Pacific from the United States when a friendly face appeared in what had been an empty seat alongside me. 'I got a lot out of reading your report on climate change', said the chairman of one of Australia's largest greenhouse gas emitters, formerly chief executive officer of another. 'But I have a question. Why did you go for an emissions trading scheme and not a carbon tax? There's going to be such a fight about free permits for trade-exposed industries because everyone can see exactly what's happening. With a carbon tax, you could just make the exemptions and everyone would forget about them, just like all the other tax exemptions.'

Well, I don't think it would have been quite like that. Apart from anything else, I saw my job as making sure that Australians understood the implications of policy decisions that were eventually taken. Every dollar of revenue from carbon pricing is collected from people, in the end mostly households, ordinary Australians. Most of the costs will eventually be passed on to ordinary Australians. Every dollar handed out for one purpose is not available for something else. Here we discuss the best uses of the carbon revenue.

The carbon price is the central element of a set of policies that will secure large reductions in Australia's emissions at the lowest cost to the Australian economy. In addition, unlike regulatory or direct action measures, a market-based mechanism can collect revenue in a way that is more efficient than some existing taxes, for use in raising productivity, promoting equity, encouraging innovation in low-emissions technology, providing incentives for sequestration in rural Australia, and easing the transition for trade-exposed industries.

Using direct action measures to achieve a similar amount of emissions reduction would raise costs much more than carbon pricing, but would not raise the revenue to offset or reduce the costs in any of these ways. The costs might be covered by budgetary expenditure, but this affects who pays the costs, not whether the costs are there. Other people's taxes have to rise to pay for expenditures under direct action.

In the long run, households will pay almost the entire carbon price as businesses pass carbon costs through to the users of their products. Various owners of business assets involved in international trade might carry part of the costs through a transition period; but, again in the long run, only business owners who earn 'rents' from natural resources or control of monopolies would have their incomes diminished.

It might seem appropriate, therefore, to pass on all revenues from carbon pricing to households as tax cuts and in other ways. But there are other claims on part of the revenue that carry larger benefits to Australians in the early years. Later, it is appropriate for the share of revenue being passed on to households to rise.

For the revenue that is passed on to households, the way in which it is applied has large implications for economic efficiency. We can substantially reduce the economic cost of reducing emissions by using the revenue from a carbon price to replace inefficient taxes.

It is sometimes suggested that providing households with assistance would cancel out the benefits of introducing a carbon price. It is said that, if we impose a carbon price that costs a household $100 and then provide that household with a tax cut worth $100, nothing has changed. These suggestions are wrong. The carbon price, even with the tax cut, alters the relative prices of more and less emissions-intensive goods and services. High-emissions goods become more expensive relative to low-emissions goods. Demand for the former falls, while demand for the latter rises. And putting a price on emissions encourages producers to use less emissions-intensive processes to produce goods and services.

For example, electricity—being relatively emissions-intensive in current circumstances—will rise relative to other prices with the introduction of a carbon price. A household facing a higher electricity bill will have an incentive to reduce its electricity consumption over time. If the household receives money through a tax cut to cushion the impact of higher electricity prices, there is no reason why it will spend all of this assistance on electricity. The household can be expected to spend the tax cut on a range of goods and services, guided by prices that take into account the costs of emissions. Regardless of the assistance, electricity will still be relatively more expensive, so electricity consumption can be expected to fall over time.

The success of a carbon price in altering the relative prices of more and less emissions-intensive goods depends crucially on the nature of the assistance provided to households. If assistance is directly linked to the consumption of relatively emissions-intensive goods (for example, rebates related to the amount of electricity used), then it will remove the incentive for the household to switch away from more emissions-intensive goods and towards less emissions-intensive goods. A tax or social security adjustment would not discourage households, now facing relative prices that reflect the socials costs of the goods they consume, from lowering their emissions.

Benefits of tax reform

A carbon price of $26 per tonne of carbon dioxide equivalent would generate around $11.5 billion in potential revenue from the value of permits in 2012–13.

The amount of revenue rises with the carbon price, but falls as emissions decrease. The revenue from a carbon price is expected to rise for a decade or so. In the longer term, the revenue from a carbon price will stabilise and then start to decline as a result of steady falls in emissions eventually overcoming the rise in permit prices.

A carbon price has some short-term negative effects on productivity growth and incomes—although less than direct action that secures similar reductions in emissions.

The modelling for the 2008 Review, and the Treasury modelling for the Rudd government's Carbon Pollution Reduction Scheme, ignored the benefits to productivity and incomes that could be secured by judicious use of the revenue from the carbon price. Carbon price revenue can be used to improve the tax system through reducing tax disincentives to work.

Other modelling has found that tax reform could offset a substantial share of the fall in rates of growth in incomes resulting from a carbon price. Analysis updated for this book shows that using carbon price revenue to fund well-designed tax reform could halve the impact on GDP of achieving the minus 5 per cent emissions reduction target in the period to 2020. Another way of looking at the results is that well-designed tax reductions allow Australia to achieve a minus 15 per cent emissions target in 2020 with around the same projected economic costs as achieving a minus 5 per cent target without reductions in income tax.

It should not be surprising that the benefits of well-designed tax reform are substantial. Economists and others have been calling for reforms along these lines for more than a decade—calls that have been echoed in the 2010 Henry tax review.

A large part of the gains in national income from tax cuts comes from increased participation in the labour force and employment. The gains extend beyond the effects on incomes: increases in employment are intrinsically valuable, enabling individuals to contribute and be valued in additional and important ways.

Existing taxes (including income tax, savings tax and indirect taxes) reduce incentives for some people to participate in the workforce. Low-income earners, for example, are typically more sensitive to tax rates than high-income earners. Decisions by mothers on whether to undertake paid work are particularly sensitive to their effective tax rate. The introduction of a carbon price to reduce emissions without a reduction in other taxes would result in less growth in real wages, thereby reducing work incentives further. Reduction of emissions to the same extent through regulatory action would reduce incentives by even more.

On the other hand, the introduction of a carbon price with a judicious reduction of other taxes may actually increase work incentives.

This suggests that there is a substantive case for linking 'revenue-positive reforms' (carbon pricing) with 'revenue-negative reforms', such as reductions in high effective marginal tax rates and associated disincentives to labour market participation.

Dividing the pie

Efficiency and equity objectives would be well served by allocating much of the revenue to reducing personal income tax rates on households at the lower end of the income distribution. This could be the kind of tax and social security reforms envisaged in the Henry review. Such an adjustment would increase incentives to participate in the labour force at a time when Australia faces shortages of labour and inflationary pressures. There can be a substantial reduction in the disincentives to work created by the interaction of taxation and the withdrawal of pensions and benefits.

Second, for those low-income households that do not stand to benefit from tax cuts, adjustments could be made to indexation arrangements for pensions and benefits that protect against disproportionate increases in the prices of particular goods and services that these households consume in unusually high proportions. Full compensation and not overcompensation should be the objective.

Third, any additional inequities would need to be corrected by targeted support for households with exceptional energy requirements for health and other reasons.

Fourth, part of the revenue should be used for firms or the carbon pricing scheme regulator to purchase carbon credits from the land sector.

Fifth, there is a case for assisting the trade-exposed industries to an extent that offsets the effects on product prices of other countries having carbon constraints that impose lower costs than Australia's.

Petrol prices

There have been considerable concerns about the distributional effects of increases in petrol prices associated with carbon pricing, particularly for those living in outer suburban and regional areas. It is not obvious how these effects can be simply compensated at reasonable transactions costs. They are actually small in relation to incomes and compared with the effects of variations in world oil prices.

In the meantime, the increase in petrol prices following the introduction of a carbon price could be offset through a one-off reduction in petrol excise, funded by other tax adjustments that had similar or larger positive effects on emissions. The cost of a one-off reduction in excise at the time of introducing carbon pricing could be covered by reform or abolition of the preferential treatment of the fringe benefits tax arrangements related to private vehicle use, and the reduction in other subsidies for fossil fuel consumption.

The fringe benefits arrangements were identified as being highly distortionary by the Henry review. Under these arrangements, the taxable value of a car's fringe benefit falls at specific intervals as the distance driven increases. This arrangement encourages more driving than would otherwise be the case and therefore increases emissions. Abolition of the concessional treatment of fringe benefits in the form of private use of corporate vehicles would pay for the initial removal of the effects of the carbon tax on petrol and diesel. If reform rather than abolition were adopted by government, as in the 2011 budget, the balance of the costs could be covered by removal or reform of other taxation arrangements that encourage the use of fossil fuels.

For the future, the smaller incremental increases in carbon prices could be compensated through additional rounds of tax cuts, when the scheme as a whole would be contributing positively to rural incomes.

Protecting the vulnerable

Protecting the most vulnerable is critical to the success of the carbon price. The reform of income tax of a kind proposed in the Henry review efficiently addresses equity concerns for most taxpayers on low and middle incomes. For households with little or no income, the transfer system provides a general social safety net. It insulates the most vulnerable from structural change to a large degree because payments rise at least in line with prices (as measured by the CPI). So even if the rest of the economy suffers a negative shock that reduces real income, the nominal levels of benefits automatically increase for the most vulnerable. However, indexation is not perfect.

Indexation may not reflect exactly the price increase that consumers face, for two reasons. First, indexation is measured on a typical basket of goods. Consumers with different levels of income consume different baskets of goods. Analysis conducted for the 2008 Review suggested that the CPI would have risen by 1.1 percentage points following the introduction of a carbon price at $23 per tonne of carbon dioxide equivalent in 2010, whereas the prices faced by one-fifth of households with the lowest incomes would have risen by 1.3 percentage points. Second, the CPI does not take into account the change in goods consumed that results from the introduction of a carbon price. As consumers are expected to switch away from relatively emissions-intensive goods (such as electricity) following the introduction of a carbon price, indexation may overstate the price rises faced by households.

Recipients of pensions and benefits face higher prices before they receive a higher payment. This is due to a lag in the availability of data and in the timeliness of adjustment. For instance, the indexation of the pension and Newstart Allowance lags behind price increases by between three and nine months and that of Youth Allowance by between six and eighteen months. It is appropriate for the government to bring forward indexation of benefits with the introduction of a carbon price, while smoothing down indexation later to avoid overcompensation. This approach was adopted when the goods and services tax was introduced.

Care needs to be taken in changing social security arrangements that there is no exacerbation of existing high marginal effective tax rates. Changes in social security and tax arrangements taken together should be designed to substantially reduce disincentives to work.

Many pensioners are a particularly vulnerable group as many are unable or reasonably disinclined to supplement their transfer payment by working. The focus here should be on preserving assistance to those on the full-rate pension. Pensions typically rise in line with wages, as a benchmark applies to ensure that they do not fall below a fixed share of male total average weekly earnings. Generally, wages rise more than prices. But in periods of high inflation, prices could rise more than wages, and so pensions increase by the greater of the two.

However, over time—when wages return to growing faster than prices— pensions will revert to the same fixed proportion of wages as they would have received in the absence of high inflation. In order to preserve their real income, compared to what it would have been in the absence of a carbon price, assistance should be delivered through a supplement, the real value of which is preserved over time through price indexation, as was the case with the introduction of the goods and services tax.

Some households use higher proportions of their income on electricity, gas and other goods and services that are particularly emissions-intensive and so experience especially large increases in costs. Some low-income households use much more electricity and gas than others, because some members have health problems or disabilities requiring special treatment. This was one reason why the government's proposals for an emissions trading scheme in 2009 provided for 'overcompensation' of low-income households.

It would be better to deal with the problem of undercompensation of households with special energy requirements directly. Households with special energy requirements can be identified through state and territory governments and private organisations, and provided with lump sums that compensate for their exceptional requirements without removing incentives to reduce energy use. This will deal with the problem more reliably, while leaving more revenue for productivity-raising taxation reform for workers on low and middle incomes.

Trade-exposed industries

The 2008 Review outlined the case for transitional assistance to emissions-intensive, trade-exposed industries. These industries have high emissions per unit of output and are highly exposed to international competition.

There are two propositions supporting this case.

First, imposing a carbon price in Australia ahead of similar carbon constraints in our trade competitors, if it were to occur, could result in some movement of emissions-intensive, trade-exposed industries from Australia to other countries that impose less of a carbon constraint. This could result in an increase in global emissions—in the event that the activity moves to a country that uses a more emissions-intensive production process than Australia. This is the universally recognised environmental risk of carbon leakage.

This risk is difficult to quantify precisely. Analyses in Australia, Europe and the United States consistently suggest that the risk is real, but exaggerated in popular discussion. We should recognise that not all movement of production from Australia to other countries would involve carbon leakage. For example, Australian aluminium production is among the most emissions-intensive in the world, as it is mainly based on coal, some of it brown coal with exceptionally high emissions. The expansion of aluminium smelting elsewhere in response to reduced smelting in Australia is likely to generate electricity from water flows or natural gas, with zero or low emissions.

Second, if Australia were to impose a cost on carbon emissions which preceded or exceeded that of countries that are the hosts to major competitors, this could cause Australian production to contract below the level that would eventuate when our competitor countries faced a similar cost. Such a loss in productive capacity would be inefficient and costly to regain at a later date when most countries were imposing carbon constraints with similar costs to Australia's.

Of course, the opposite propositions are equally true and just as important for economic efficiency when Australian action lags behind that of competitors. If Australia falls behind other countries on mitigation, there will be incentives for uneconomic expansion of the favoured industries. This, in turn, damages other industries through the effects on interest and exchange rates and costs. There is as much damage to the economy in over-assistance as in under-assistance.

Accepting the two propositions that argue for positive assistance suggests a number of design features that will need to be in place when carbon pricing is introduced.

First, assistance will be of a transitional nature pending comparable carbon pricing in the rest of the world. Second, assistance should only compensate for the inefficient distortion arising from an uncoordinated global start to emissions reduction, with sales prices for emissions-intensive goods being lower than they would be if all countries imposed similar carbon restraints to Australia.

This means that assistance to all firms should be withdrawn once most countries are imposing similar carbon constraints. Some countries may continue to assist specific sectors and to create distortions even after most countries are imposing similar constraints. Such counter-subsidising would contribute to a destructive, reinforcing cycle of protectionism. It is important for Australia to work with other countries to secure international application of sound principles to avoid continuing distortion.

The 2008 Review described an approach to assistance based on avoiding the transfer out of Australia of production that would remain if other countries imposed similar carbon constraints to Australia. I called this the 'principled approach'.

While it would be desirable to move promptly to the principled approach to assistance for trade-exposed industries, this is not practical, as it will take some time to put in place arrangements to administer the scheme. The arrangements proposed for trade-exposed industries within the government's 2009 Carbon Pollution Reduction Scheme could be applied for the first three years, while institutional arrangements are established for the principled approach. In these three years, the 'buffer' for the effects of the global financial crisis should be recovered, as it has been made redundant by recovery. The implementation of the interim arrangements almost certainly provides excessive assistance to some industries. This is especially unfortunate at a time when subsidising incomes and employment in one sector forces reduction in incomes and employment in other industries that are under stress from the resources boom (see Chapter 7). It may also provide under-assistance to some industries. Different observers will have different views on whether over-assistance exceeds under-assistance. These differences would be resolved through the work of an independent agency.

The pressure that is being applied to other Australian industries by the resources boom makes any over-assistance to the resources sector especially unfortunate at this time. The revision of assistance under the principled approach for the trade-exposed industries within the resources sector at the end of the three-year interim period is a matter of great importance and priority.

An independent agency should be responsible for developing the approach to emissions-intensive, trade-exposed industry assistance beyond the first three years of the scheme. The agency would have features similar to the Productivity Commission and could be the Productivity Commission.

The agency should be asked to review the new approach, and to vary it in the light of analysis and experience if variations would raise the incomes and welfare of Australians. It should develop a suitable work program to ensure priority sectors are considered early, in anticipation of the switch to the new, principled approach. Priority should be given to data collection and analysis on emissions-intensive, trade-exposed industries, which are receiving the largest amount of assistance.

Once a move from the interim to the new approach has been made, the agency should continue to provide advice on the operation of the assistance regime, including advice on when global carbon pricing has progressed to the point where there is no longer an economic justification for emissions-intensive, trade-exposed industry assistance for Australian firms.

The independent agency would be backed with the necessary resources and would have the professional capacity to do this job well. It would operate transparently in the manner of the Productivity Commission, exposing its methodology and data sources for public comment.

Assistance provided to emissions-intensive, trade-exposed industries to correct for undesirable and inefficient outcomes should not be confused with providing support to industry for the loss of profits or asset value arising from the introduction of a carbon price in Australia. Any fall in asset value stemming from the change in relative pricing creates no greater case for compensation than other government reforms to reduce other market failures. The introduction of measures to discourage smoking, to control the use of asbestos, to raise occupational health and safety and environmental standards, and to reduce lead in petrol are all cases in point.

The land sector

For good reason, agriculture and the land sector will not be comprehensively covered by carbon pricing in the early years. There are large advantages in allowing genuine sequestration in the land sector to be rewarded at the carbon price, whether or not that is currently allowed under the international rules developed at Kyoto and currently under discussion with a view to modification. There is great uncertainty about the claims that the land sector may make on carbon revenue, but they are potentially large. Chapter 10 suggests that, pending full coverage of the land sector in carbon pricing, provision be made for a proportion of the carbon revenue to be allocated for land sector credits.

Innovation

Public funding of low-emissions innovation over the medium term is necessary to compensate for the external benefits deriving from a private firm's investment in innovation, at a time when there is a high value in accelerated development of new, low-emissions technologies.

Chapter 9 explains the case for public funding of innovation in low-emissions technologies to rise to about $2.5 billion a year for policies across the innovation chain. The government is currently allocating about three-quarters of a billion dollars a year to innovation in low-emissions technologies through the three-year forward estimates and beyond. This funding will presumably continue, so that the carbon pricing package has to fund only the increase above three-quarters of a billion.

Table 6.1 brings together the recommended uses of the revenue in a budget-neutral framework.

Table 6.1: How it fits together[a]

	Fixed 2012–13 (%)	Floating 2015–16 (%)	2021–22 (%)	Total
Total permit revenue[b]	100	100	100	100
Household assistance	55	60	60–65	60
Tax reform	40	45	50	45
Benefits payments[c]	15	15	10–15	15
Energy efficiency	1	0	0	<1
Business assistance	35	25	20	25–30
Industry assistance[d]	30	25	20	26
Electricity transition	3	0	0	<1
Structural adjustment	2	0	0	<1
Innovation[e]	10	15–20	20	15
Carbon farming[f]	5–10	10–15	15	10
Gross expenditure	105	110–15	115	110
Less market offsets and existing innovation expenditure	5–10	10–15	15	10
Net budget impact	0	0	0	0

a. Fuel reform and aid monies are not drawn from permit revenue and are therefore not shown.

b. Includes the increase in revenues for the first three years from auctioning of permits for use at later dates. The sum of percentages may not add up to 100 per cent due to rounding.

c. Around half the welfare payments under the former Carbon Pollution Reduction Scheme package.

d. Proportion of assistance paid to emissions-intensive, trade-exposed industries assumed to fall by 1.5 percentage points each year after year three.

e. This percentage includes existing innovation funding.

f. This percentage includes Kyoto Protocol offsets sold to liable entities (which do not represent a cost to the government).

Conclusion

The carbon pricing scheme will generate large amounts of revenue—about 20 per cent of that collected by the goods and services tax. With careful use, the revenue can fully compensate low- and middle-income earners for the costs of the scheme while supporting a substantial efficiency-improving tax reform. At the same time, it can support assistance to avoid uneconomic reduction of production in emissions-intensive industries, incentives for large-scale utilisation of opportunities for biosequestration in the land sector, and fiscal incentives for innovation in low-emissions technologies. Some funds would be made available for structural adjustment if it emerged that there were regions in which economic activity and employment were hit heavily by carbon pricing. Provision would be made for the possibility that loan guarantees to secure energy security were called through the early years of transition to a low-emissions economy.

The total potential revenue would include proceeds of sales of about 10 per cent of one year's permits in each year, for use at any time from three years after issue. This would support the emergence of a forward market in deliverable permits. It would increase net revenues from the sale of permits in the first three years of the scheme.

Over time, tax cuts for households, innovation and land sequestration would draw gradually more deeply on the revenues. Investing carbon revenue in these ways would boost economic growth and the resilience of our economy overall. Requirements for assistance to trade-exposed industries would gradually fall, as other countries' carbon constraints tightened, and were taken into account more accurately in Australian arrangements.

7 The best of times

IN 1985, the first big outing for the newly formed Business Council of Australia was the National Taxation Summit. The managing directors of two of our grand old companies, Westpac and WMC, represented the organisation at the summit. The Business Council and its predecessors had been seeking a tax switch for many years—the introduction of a tax on the consumption of all goods and services—to pay for a reduction of taxes that they didn't like. Prime Minister Bob Hawke had called a National Taxation Summit after the 1984 election, and this was the new council's big chance.

Officials of the Treasury and the Department of the Prime Minister and Cabinet had prepared a white paper for presentation to the summit. It included three possibilities, two of which were carefully worked out live options. Option A included the introduction of taxation on a range of activities that the political system had shielded for many years, and hence had increased the burden of taxation on everyone else. These included a tax on fringe benefits, a tax on capital gains, limits on deductions or negative gearing of property, and extension of the normal company tax to income from gold mining (which had been temporarily excluded from taxation in 1922). Option C included all of option A, plus a broad-based retail sales tax to replace the wholesale tax and to help fund large reductions in income tax.

The prime minister invited the meeting to consider carefully the relative merits of the three options before it.

Bob White, the president of the Business Council, was then called to the despatch table of Parliament House by the prime minister. He rose to his feet amid great expectation.

'The Business Council', he said, 'does not support approaches A, B or C.'

A hush fell over the meeting.

White went on to say that option C was unacceptable because it contained option A, but no-one was listening.

The hush turned to murmuring, and then loud voices talking over each other.

History had been made in a moment.

The consumption tax was knocked unconscious, and was not revived for two decades. The various components of option A were legislated progressively from 1985.

What emerged that day in Parliament House was a lesson in how vested interests can make the perfect the enemy of the good. In overreaching for an ideal outcome for themselves if not for the community, the Business Council destroyed a central pillar of tax reform for two decades. They shot themselves squarely in the foot, with the country as collateral damage.

Fast forward to the late 1990s, and the Business Council played a constructive role in building support for a tax reform package that included a consumption tax, by advocating some measures that were important for equity and valued by other groups in the community much more than by Business Council members themselves. The Australian business leadership of that new day dealt itself into serious discussion of policy reform by adopting a national rather than sectional focus.

But fast forward again to 2011 and we find a new generation of the Business Council has reverted to its old type. In April, the chairman of the Business Council, Graham Bradley, visited China alongside a visit by the Australian prime minister. During high-level discussions with senior government and business leaders, Mr Bradley said that the Business Council would not support any carbon tax that would 'discourage investment' in Australia. And there should be no carbon tax on natural gas.

There can be no carbon pricing without structural change. Structural change removes some jobs and discourages some investment. It is not logical to be in favour of a market-based mechanism for reducing emissions, as the Business Council professes to be, and simultaneously be against a carbon price that discourages any investment. It would be as illogical as favouring productivity-raising reform but being against any policy change that discourages any investment.

The proposed carbon price, leading to an emissions trading scheme, is the central element in a set of policies that will secure reductions in Australia's emissions at low cost to the Australian economy. In addition, unlike regulatory measures, a market-based mechanism can collect revenue in a way that is more efficient than some existing taxes, for use in raising productivity or promoting equity.

Using regulatory measures to achieve a similar amount of emissions reductions would raise costs but would not raise the revenue to offset the increased costs. There would be no revenue to remove distorting taxes and offset the regressive effect on income distribution. There would be no revenue to support innovation in low-emissions technologies. There would be no revenue to provide incentives for carbon sequestration in rural Australia. There would be no revenue to support trade-exposed industries even though there may be a public interest case for some assistance.

In reaching for an apparently perfect mitigation solution that appears to have no losers among its members, the Business Council is again making the apparently perfect the enemy of the good. It is elevating the cause of narrow business interests above the many benefits that add up to the national interest.

Hard times for some

While there are contradictions in the resistance in practice of the Business Council to a market-based carbon price, for some of its members there are no contradictions. The most prominent of these is the manufacturer BlueScope Steel. In April 2011, Graeme Kraehe, chairman of BlueScope, said he could not accept a price on carbon. Mr Kraehe's views found an unlikely advocate in his traditional nemesis, Paul Howes, the head of the Australian Workers' Union, whose members work for BlueScope. Mr Howes joined Mr Kraehe in an attack on the carbon price by declaring that his union would withdraw support for the carbon tax if it meant the loss of 'one job'.

To fathom the reality behind these comments, we need to understand the broader economic moment. The Governor of the Reserve Bank of Australia, Glenn Stevens, noted in February 2011 that the high prices for Australia's resource exports meant that other industries had to invest and produce less: 'On this occasion, the nominal exchange rate has responded strongly', he said. 'This ... gives price signals to the production sector for labour and capital to shift to the areas of higher return.'

He went on to say that:

> there is going to be a non-trivial degree of structural change in the economy. This is already occurring, but if relative prices stay anywhere near their current configuration surely there will be a good deal more such change in the future ... [I]f we have to face structural adjustment, it is infinitely preferable to be doing it in a period in which overall income is rising strongly. If nothing else, in such an environment the gainers can compensate the losers more easily.

In other words, Australia is enjoying a resources boom and for each new coal mine or gas plant that opens up, there must be a cut in jobs and investment in some combination of tourist hotels and restaurants, universities, steel mills, farms and other businesses producing exports or competing with imports. If it is a big investment in gas and coal, a lot of jobs and investment have to go. Prop up jobs in one area, and even more have to go in others.

The Reserve Bank's mechanism for this adjustment is higher interest rates and with them a higher exchange rate. Both will rise until enough investment has been discouraged and enough jobs have been shed in other businesses to make room for the resources boom without generating inflation.

BlueScope is one of those firms caught in this larger market shift. Its profitability is suffering enormously under the strains described by Mr Stevens. The Australian Treasury has demonstrated just how small the carbon-related changes are compared with the effects of the resources boom. With a carbon price of $20 per tonne and assistance arrangements under the government's 2009 proposals, carbon pricing would add on average $2.60 per tonne to the cost of making hot rolled steel from iron ore. This compares to current prices of around $900 per tonne of hot rolled steel. The 7 per cent appreciation of the Australian dollar in 2011 alone (a small proportion of the appreciation since the resources boom began) has subtracted about $50 per tonne from the value of steel sales. Bluescope Steel responded that this ignored indirect carbon costs and the rising price of carbon, but these do not change the basic story in the early years before the introduction of a principled approach to assistance.

When we tally these forces, it is clear that the public utterances and allegiance of BlueScope's capital and labour champions is a claim for protection against the slings and arrows of outrageous fortune, not any soundly based concern arising from the carbon price.

But while it is a good time economically for structural change, the story of BlueScope shows that it is a difficult time politically. The resources boom and full employment are forcing people out of old jobs, and it is easy to blame this on carbon pricing. It is easy to make dramatic claims about jobs being lost in one industry, knowing that not many people will understand that this means that more jobs can be created in other industries.

The introduction of measures to reduce greenhouse gas emissions also causes structural change, albeit on a smaller scale than the resources boom. But at a time when jobs are being lost, so that more can be created elsewhere, the carbon price is a ready scapegoat for those seeking to duck the consequences of the resources boom.

Hard times in resources?

Another industry that joined the 2011 self-interested hue and cry against a market-based carbon tax is mining. In early May, the chairman of BHP Billiton, Jacques Nasser, gave a speech at the Melbourne Mining Club in which he expressed general but not specific support for a market-based solution to carbon and energy management: 'It is difficult to predict winners and losers, with subsidies generally causing distortions as opposed to success.'

Yet, in the press conference following the speech, Mr Nasser also emphasised that his preferred method of reform was a slow 'sectoral approach', beginning with the electricity sector. In short, Mr Nasser was happy to endorse market-based approaches as long as they did not include coal and gas, two large components of BHP's business that would be heavily affected by carbon pricing.

Department of Climate Change and Energy Efficiency estimates provide some perspective on the coal industry's claims of impending ruin as a result of a carbon price. Carbon pricing is estimated to add an average of $2.80 per tonne to the cost of metallurgical coal ($6.70 for particularly gassy mines). But let's not forget that the price of metallurgical coal has varied between $100 and $400 per tonne during the resources boom.

The resources boom has been driven by Northeast Asian economies' growing need for energy, especially coal and gas, and other resources. Demand for gas is magnified by environmental including greenhouse considerations in Japan, China, Korea and Taiwan. So other countries' participation in the global mitigation effort has enhanced the resources boom to some extent—that is, through raised export prices, increased sales volumes and new investment in productive capacity.

The increased demand for gas in Northeast Asia, in turn, has increased the average incomes of Australians. But it has also increased the real exchange rate and reduced the availability of capital to other industries. It has therefore increased pressures for structural adjustment in other parts of the economy. As well, it has increased the difficulty and the costs of Australia meeting the emissions reductions targets that are our fair share of the international effort.

As we saw in Chapter 2, the resources industries, especially gas and coal, are responsible for around half of the extraordinary increase in Australian emissions that is anticipated in the absence of strong mitigation policies. If baseline emissions in 2020 are going to be 24 per cent above 2000 levels as supposed by the Australian Government—and that estimate from late 2010 preceded announcements on several gas export projects—then Australians will have to pay for the extra entitlements or reduce emissions elsewhere.

Despite these realities, the liquefied gas export industry is demanding exemption from a carbon price. It has based this on the claim that the beneficial effects of gas exports in reducing emissions in other countries mean that the gas industry should be exempted from responsibility for its emissions within Australia. Australia still has to meet its targets if emissions rise because of the expansion of gas and coal production.

The gas export industry is richly rewarded for the beneficial effects of its emissions being low compared with coal (although high compared with energy sources that do not rely on fossil fuels). It is rewarded in the strong demand and high prices that are driving the current investment and export boom. Some Australians have to pay for the gas industry's emissions. But why should all Australians carry the costs of the gas industry's exceptional expansion and prosperity? Why should the education, farming, tourism and manufacturing industries pay for the extra emissions that have come with the exceptional prosperity of the coal and gas industries, when their own prospects have been damaged by the resources boom?

Australian gas and coal have been the beneficiaries of a once-in-a century (or, more accurately, once-in-history) boom in demand for energy but want other Australians to pay for the resulting emissions.

Reform in the public interest is impossible in these circumstances unless there is an informed centre of our political community that understands the issues and has no sectional interest that leads it to oppose the national interest. If the independent centre of our political community is to embrace carbon pricing, it must also understand the resources boom and the 21st century collapse of productivity growth in Australia.

Booming incomes and slumping productivity

Australians are enjoying the best of times in our material standards of living. Over the past two decades we have enjoyed the longest period of rising living standards unbroken by recession in history—our own, or that of any other developed country.

We are back near the top of the world's league table for average incomes. For the first time in a century, average incomes in Australia rose above those of the United States in early 2008 (see Figure 7.1). Australian incomes have bounded ahead since then as the Great Crash of 2008 sent the US economy into the doldrums and left Australia relatively unscathed. This long boom began with, and was supported by, Australian economic reforms from 1983 to the end of the 20th century. These reforms lifted our productivity growth relative to that of the rest of the developed world higher in the 1990s than it had been since our Federation. Much higher. For a while we were at the top of the productivity growth world table of developed countries, after being at the bottom with New Zealand, on average, for over eight decades.

The lift in productivity came from improvements in the efficiency with which we used all resources—labour and capital together—and not just from using more and more capital with each worker. This increase in total productivity was the basis for sustainable increases in living standards.

As we entered the new millennium, and zeal for reform faded, so too did the surge in productivity (see Figure 7.2).

**Figure 7.1: Labour productivity and gross national income per capita
(Australia as a percentage of the United States)**

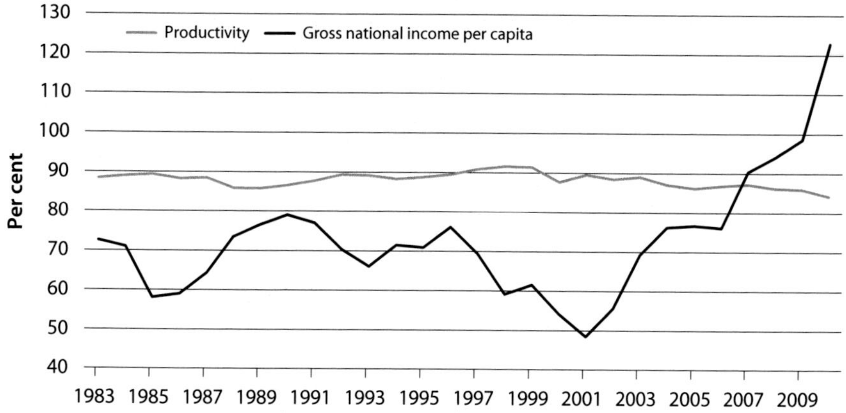

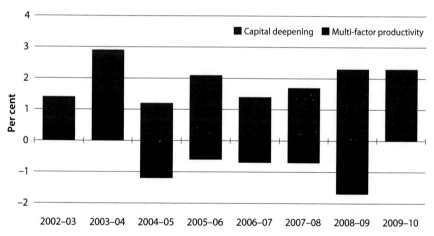

Sources: US Bureau of Economic Analysis 2011; Australian Bureau of Statistics 2011; Conference Board Total Economy Database, September 2010.

Figure 7.2: Sources of higher labour productivity

Source: Australian Bureau of Statistics 2010, *Experimental estimates of industry multifactor productivity, Australia: detailed productivity estimates*, cat. no. 5260.0.55.002.

For some years after 2001 we boosted incomes unsustainably through a rise in offshore borrowing for housing and consumption. That boom would have ended in tears except for the timely arrival of the largest sustained jump in our terms of trade in our long history as an exporter of commodities. It still would have ended in tears with the Great Crash, if the federal government had not taken the unprecedented action of guaranteeing the banks' wholesale debt—eventually to the tune of around $170 billion. The high terms of trade were then quickly restored, and continue today. It helped that we did not have a recession during and after the Great Crash—a product of superior institutions as a result of late 20th century reform, as well as deft policy at home and in China.

It is impossible to overstate the significance for productivity growth and future economic performance of the reversion to pre-reform Australian political culture that came in the early 21st century. In such a culture, economic reform is impossible if there is any prospect of there being a loser, no matter how large the gains for the community as a whole. In such a culture, economic reform is impossible if it requires any restraint on present incomes, no matter how large the benefits for the future. Productivity-raising economic reform— reform to protect society against future losses in productivity—is therefore impossible.

The future prosperity of Australians depends on us now breaking this great Australian complacency of the early 21st century.

Boom, bust and carbon

In the mining business, the ideal is for the highest economic value ('rent value') resources to be developed first and economically marginal resources last. Australia has its share of highly valuable mineral and energy deposits, so current high prices are driving a high level of investment in Australia.

Two other factors also encourage the current boom. Australia's fiscal regime for the resources industries place lower burdens on marginal investments than those placed on competing suppliers of energy and metallic minerals to international markets. In addition, sovereign risk is lower in Australia than in competing exporters of resources. This means that some poorer Australian resources jumped the queue and were mined relatively early. Some better resources in developing countries were held back while uncertainties in policy and national governance were resolved. But mined they will be, leading to a time when developing countries will be host to a higher and Australia host to a lower share of global resources investment.

Resources booms don't last forever. Eventually, high prices encourage investment in many resource-rich countries and not only in Australia. Prices for commodities fall. The growth in new global investment in mines and then the level of investment fall—and most of all in countries like Australia, which enjoyed more of the early boom. There is downward pressure on the exchange rate and incomes.

The economically wise approach to managing such episodes is to ensure that national savings are especially high through the boom, so that expenditure can be maintained in the subsequent slump in incomes. This requires higher collections of taxation revenue, higher budget surpluses in the boom times, and wise investment of the surpluses outside the domestic economy.

Restoring productivity growth

Reducing emissions through carbon pricing has a small negative effect on productivity for a while. But the alternative to an efficient approach to reducing emissions through carbon pricing isn't to take no action at all. It is rather the adoption of jerky regulatory interventions, one after another. Each will become more costly and intrusive as governments react to concern that Australia is nowhere near approaching targets for reducing greenhouse gas emissions that have been agreed with the international community, and to persistent electoral pressure. The experience with regulatory interventions so far in Australia and elsewhere is that their cost is many times the cost of securing similar emissions outcomes through general carbon pricing.

The threat that the 21st century return of the anti-productive Australian political culture will be longlasting is much greater if regulatory approaches are taken to reaching emissions reduction targets. The opportunities for vested interests to influence the policy process are much greater because the government must negotiate individual solutions to mitigation challenges as they arise. The difficulties of establishing a basis for international trade in entitlements are greater. The technical difficulties of assessing assistance levels through objective and independent processes are greater. And the danger that vested interests in other countries will persuade their governments to punish Australia for not doing its fair share in mitigation is greater.

The largest cost of mitigation through regulation is the damage that it will do to productivity-raising reform. Expansion of regulatory intervention will entrench the pressure of vested interests on the political process and the anti-productive political culture of the early 21st century.

Strong productivity and flexible markets are the cushions upon which the eventual bust will fall. We need, therefore, to be mindful of the choices for mitigation that we make during the boom so that we do not ultimately make the bust far worse.

The central choice is about which policy instrument we should use to reduce greenhouse gas emissions—carbon pricing or direct action? Beyond that central choice, future Australian productivity will be greatly affected by choices of rules for assistance for trade-exposed industries, and for trade in emissions entitlements.

It is essential to move quickly to place assistance for trade-exposed industries on a principled basis. Entitlements to assistance must soon be determined by a credible, independent and well-resourced institution, applying transparent analysis based on clear principles derived from analysis of the national interest.

The principled approach would be to provide assistance to the extent that product prices would be higher if all countries had Australian carbon constraints. Logically, there would be a levy in industries in which carbon constraints elsewhere exceeded those in Australia, but that would be a bridge too far.

The disciplines imposed by the Tariff Board and its successors provided important support for the emergence of a political culture in Australia in the 1980s in which productivity-raising reform became possible. Similar disciplines are going to be important to protect climate change mitigation from old Australian patterns of resistance to necessary structural change.

Second, Australia should move strongly to establish frameworks for legitimate international trade in entitlements. Trade could be conducted first on bilateral and regional bases, which can link as soon as possible to mitigation efforts in other countries and regions. This will provide early opportunities for deep international trade in emissions entitlements. Trade in entitlements will lead to convergence over time in carbon prices, which removes arguments for assistance for trade-exposed industries.

Not all countries will be open to deep participation in trade in entitlements, even if they are making strong steps to reduce emissions at home. The United States may be such a country for a considerable time. This will delay the emergence of a truly global carbon price. This, in turn, will increase the risk of distortive interventions to enhance the position of trade-exposed industries. Beyond the damage that this will do to the integrity of national policy making, there is a serious danger of a breakdown of the rules-based international trading system. It is possible that Australia and New Zealand would be damaged more than any other countries by such a breakdown, unless it were clear to a prejudiced observer that they were doing their fair shares in the global mitigation effort.

Finally, there are several ways in which economically efficient reduction in greenhouse gas emissions may positively contribute to the end of the stagnation of Australia's productivity. Concerns highlighted in the debate over carbon pricing have opened the way for reform of electricity price regulation, with potentially significant effects on the productivity of capital use. Incentives for biosequestration may accelerate the use of farm management approaches that raise productivity, including through reducing vulnerability to drought. The use of carbon revenues for tax reform could increase efficiency in the labour market.

Conclusion

An emissions trading scheme, initially with a fixed price on carbon, will be introduced at a time of great prosperity in Australia—a time of full employment but also of emerging structural pressures from the resources boom. With Australia's exchange rate against the US dollar at its highest level in about 30 years and its real (inflation-adjusted) exchange rate possibly the highest since Federation, we are living through the largest reallocation of resources outside the two world wars in our national history. Developing countries' accelerated global industrial development will drive and restructure the Australian economy in the years ahead. There will be bumps in the road—but these will probably be less painful than they would have been in any other circumstances. And the bumps will be on a road that is, for the foreseeable future, heading in directions that are favourable for Australia, determined by the concentration of global growth in economies that are highly complementary to Australia, and in our neighbourhood.

It makes no sense to resist this change with policies that seek to hold in place the structures of the past. On the other hand, it makes good sense to ensure that policies pursuing different objectives are all consistent with continued increases in productivity and rising living standards after the current resources boom has run its course. That means adopting approaches to reductions in greenhouse gas emissions that have the lowest possible costs. It means using the revenues from carbon pricing for tax reform to increase labour force participation and productivity at the same time as we meet important equity goals.

Adapting to the inevitable climate change in the remainder of this century requires the same efficient markets and flexible economic structures that will be necessary for efficient reduction of emissions and for restoring productivity growth as we live through the continued rise—and then the fall—of the resources boom.

The old Australian political culture, which was resistant to structural change, and which responded to private and sectional rather than public and national interests, is inimical to success in this historic national challenge.

8 Adapting efficiently

THE BEST of mitigation will leave Australians dealing with a lot of climate change.

They will have no choice but to adapt.

But to what will they be adapting? While the climate outcomes from the Cancun Agreements cannot be defined even in broad brush because they say nothing much about what happens after 2020, Chapter 4 suggests that they could lead to atmospheric concentrations of greenhouse gases of 550 or 650 parts per million—most likely leading to temperature increases of 3°C or 4°C. It is still possible that the Cancun pledges could evolve into a set of commitments that achieves the Cancun temperature objective of below 2°C. And it is not impossible that future Australians could face an increase in global temperature of 6°C or more.

The range of uncertainties is wide and extends into territory in which it is unrealistic to think that a national policy response can be coherent or even relevant. Beyond a certain point government would be overwhelmed by the impacts of climate change.

We are already feeling some impacts of climate change when the increase so far is less than 1°C since pre-industrial times. How will Australians in future manage 2°C, which for the moment seems a lower bound on a wide range of possibilities?

Even an increase of 2°C above pre-industrial levels would have significant implications for the distribution of rainfall in Australia, the frequency and intensity of flood and drought, the intensity of cyclones and the intensity and frequency of conditions for catastrophic bushfires.

The difference between 2°C and 3°C was examined in detail in the 2008 Review. It is large. And every degree upwards after that is worse. There is no point at which we can say that so much damage has been done that there is not much point in stopping more.

Let us say that the International Energy Agency is right and that in the absence of a decisive change in policies we are headed towards the atmospheric concentrations of greenhouse gases that would give us a temperature rise of around 4°C.

A global average temperature rise of 4°C from pre-industrial levels (3.5°C above 1990 levels) is well outside the relatively stable temperatures of the last 10,000 years, which have provided the environmental context for the development of human civilisation. We would be in unknown territory for humanity.

A temperature increase of 4°C above pre-industrial levels would give an 85 per cent probability of initiating large-scale melting of the Greenland icesheet, put 48 per cent of species at risk of extinction, and place 90 per cent of coral reefs above critical limits for bleaching. It would trigger the lower threshold for initiating accelerated disintegration of the west Antarctic icesheet and changes to the variability of the El Niño – Southern Oscillation, and the upper threshold for terrestrial sinks such as the Amazon rainforest becoming sources of carbon rather than sinks.

There are two main building blocks for a productive response to the adaptation challenge. The first is to make sure we have a strong, flexible economy, with smoothly functioning markets. The second is to make sure we have sound information about possible impacts of climate change on various regions and activities and that information is disseminated in easily useable forms.

These are the most valuable things that we could bequeath those who come after us as they do their best in a world of climate change. Adaptation policy is first of all about doing these things well.

A resilient and flexible economy

It is an obvious point, but true, that the high probability of dangerous climate change strengthens the reasons for Australia making sure that it has a strong and flexible economy based on a well-educated and adaptive people.

Climate change strengthens the importance of Australia quickly getting back onto a path of strong productivity growth, built on efficient markets and effective economic policy-making institutions that are able to define and implement policy in the national interest.

There will be shocks and hard times, some coming from the direct effects of climate change on us, and others from the effects on other countries that are important to us. Australians in future will do better if they are working with a productive economy, which is in a strong fiscal position in preparation for a shock, and has the structural flexibility that comes from well-regulated markets.

These strengths are the less likely to be tested beyond their limits the more effective global action has been in constraining climate change.

So current mitigation policy is an important foundation for future adaptation policy. And, similarly, adaptation options should be designed with an awareness of their impact on mitigation policies.

The challenge of future climate change makes it even more important to minimise the costs of mitigation. Doing our fair share in global mitigation will have a cost—and in the early years a net cost before the benefits of avoided climate change are brought to account. It is important that this cost is the lowest that it can be. A similar argument applies to adaptation.

Here the advantages of carbon pricing over regulatory or direct action are twofold. First, the immediate and direct sacrifice of some productivity growth for mitigation will be much smaller if a carbon price encourages millions of Australians to find, and sometimes to invent, ways of reducing emissions at lowest cost, rather than having a few political leaders and their advisers and close associates identifying clever ideas for direct action. Second, and of fundamental importance, the many interventions involved in making large reductions in emissions through direct action would encourage the return to the old-style Australian political economy. When we need to remove the great Australian complacency of the early 21st century, a regulatory approach to mitigation would entrench and extend it.

Adapting through markets

As with reductions in emissions, adaptation to climate change will be more effective and secured at lower cost the more individual Australians and enterprises as well as governments at all levels are involved in working through the choices, anticipating problems before they arrive and taking into account all of the risks in their investment decisions.

Soundly functioning markets assist households, communities and businesses to respond effectively to the impacts of climate change. Markets provide the most immediate and well-established avenue for addressing many of the uncertainties posed by climate change.

Australia's prime asset in responding to the adaptation and mitigation challenges that lie ahead is the prosperous, open and flexible market-oriented economy that has emerged from reform over the last quarter century. Government can facilitate adaptation by continuing to promote broad and flexible markets, and seeking to correct remaining barriers to their efficient operation.

Some domestic and international markets for particular goods and services will be especially important to Australia's adaptation response. These markets may require increased policy attention to remove barriers that limit the ability of markets to harness efficient adaptation. Included in this category are markets for insurance and finance, water and food.

Insurance and financial markets

Households and businesses are able to manage many risks effectively through the insurance and financial markets. As the frequency and intensity of severe weather events increase with climate change, demand will rise for related insurance and financial services.

The recent innovation and deepening in insurance markets shows their considerable potential to promote adaptation to climate change. By its nature, however, conventional insurance is of limited value when an adverse event is likely to have similar impacts over wide areas of the world. Nor is conventional property insurance of much help when the uncertainty mainly involves the timing rather than the extent of an impact.

An example is sea-level rise if it were to become clear that the melting of the Greenland icesheet had become irreversible. It would then be inevitable that large numbers of coastal properties would be inundated, but uncertainty would remain about the timing of the loss. There might then be scope for developing new property insurance products that share characteristics with traditional life insurance. Life insurance covers the risk of timing of death, although the fact of eventual death is itself certain. The development of innovative products that matured on loss of property and that would provide the means of buying housing elsewhere if the insured event occurred may be seen as having value and could be developed by the commercial insurance sector. The commercial viability of such instruments would depend on insurance companies being able to develop a balanced portfolio of insurance and financial risks in a world of climate change.

The expansion and dissemination of knowledge from applied climate change science can assist the development of new insurance products for these circumstances. To the extent that state and local government decisions about land-use planning and zoning are based on sound knowledge from the climate science, there will be improvements in the operation of relevant insurance markets.

As the Henry tax review has noted, insurance products are subject to a range of insurance transaction taxes and direct contributions to the funding of fire services, which leads to inefficient outcomes. The interaction of these taxes and levies increases the cost of premiums, which may reduce insurance uptake. The revenue benefits of such taxes need to be evaluated against their inefficiencies and economic costs, particularly given the role of insurance in encouraging firms and households to adapt to climate impacts.

Water markets

The challenges for rural and urban water supply result from the interaction of climate change with increased demand from growth in population and economic activity. The limited scope of markets has complicated the task of allocating water to its most valuable uses. Chapter 10 notes that there will at times be local reasons for constraining landowners' decisions on the uses to which their land and water assets can be put. But these limits should only be applied when there is good reason to do so, and land-use planning should generally be directed by affected communities.

The same rules should be applied to water use. There are advantages in water being covered by property rights and regulated for sustainability, and for the owners of those rights to be able to apply the water to uses of their choice unless there are good local land-use planning reasons to constrain private decisions.

Australia's rural water market is the result of many years of reform, but some barriers to efficient operation remain. While extraction of in-stream flows has been regulated and subsequently subject to a price, access to groundwater and surface flow has often been left as a common property resource, with predictable consequences.

The 2008 Review noted that the establishment of a well-functioning water market that delivers the best possible outcomes in the context of climate change will require the active involvement of government. Government is required to establish the most effective administrative and regulatory arrangements for the functioning of the market. Once the water market has matured, the role of government moves to one of adequate monitoring and enforcement.

But barriers to efficient water management in a changing climate persist. For example, in water markets, regional restrictions on trading remain a significant barrier. Severe water shortages in urban centres have led to the development of a number of desalination plants in Australia over the past few years, at high cost. The Productivity Commission has questioned the cost-effectiveness of some of this expenditure. Would wider market exchange of water, with desalination plants competing with bids from a range of sources including long-distance storage, have produced a good result at lower cost?

In the nature of market exchange, we would only find out by trying it, but the general experience is that market processes often generate results that are surprisingly good.

Food markets

In the absence of effective and ambitious global action, deep participation in international trade in food as an importer as well as an exporter is going to be important for Australian food security. This is going to require the easing of inhibitions about the import of food. This will be stressful for many rural Australians in particular, but the alternatives will be worse. The importance of free trade in food to food security in a world in which there has not been effective and strong mitigation is discussed in Chapter 10.

An informed Australian people

Sound information is the second foundation for effective adaptation to climate change. Informed people and enterprises and governments at all levels will see problems in advance and develop low-cost responses to them. On the other hand, people and firms and governments responding to crisis will make decisions without the benefit of long reflection and consideration of alternatives to what the crisis seems to demand.

Here I should draw attention to another of the costs of so-called 'scepticism' about climate change science beyond its interference with the development of sound mitigation policies. If a proportion of Australians are persuaded that the mainstream science is wrong or unreliable then they are denied information that is essential to the exercise of sound judgments about many decisions that affect the quality and cost of adaptation.

As the average rainfall declines sharply with each passing decade in the south-west of Australia, a farmer who shares the scientific knowledge that is the common heritage of humanity will make different decisions about land use than one who thinks that a series of dry autumns is a passing phase. The regulators of power distribution in a state that has just been devastated by a bushfire during what would once have been described as once-in-a-century conditions will make different decisions if they know from science that these conditions will now arrive with awful frequency.

Improvement of applied climate science and dissemination of the outcomes will not assist adaptation decisions by those who have closed their minds to uncomfortable reality. As is the case with denial of science in many areas—Professor Peter Doherty asks us to consider denial on immunisation and transmission of AIDS as parallels to climate science denial—the isolation of some people from reality can damage the adaptive response for others in the community.

In any case, we need more and better information on the likely impacts of climate change on various parts of Australia, and we need that information to be readily available for those who require it for decisions on many things. There are several aspects of the applied climate science that work out differently here than in the northern hemisphere. As the leading country of science in our hemisphere, we will have to do a lot of the science ourselves.

We have made progress on building our national strengths in climate change science since the 2008 Review noted that our capacities in this area, while of high quality, were inadequate to the national task. I observed in 2008 that while pluralism in science was desirable in itself, the importance of scale for some of the large modelling tasks in particular meant that integration of the national effort was important.

The joint CSIRO – Bureau of Meteorology Centre for Australian Weather and Climate Research has strengthened Australia's capacities as it was intended to do. The National Climate Change Science Framework, the National Climate Change Adaptation Research Facility and the CSIRO Climate Adaptation Flagship are making substantial contributions. This work is of great importance for effective adaptation to a changing climate.

The dissemination of the results from applied science to the people who are interested and who would make use of them is an important task. Also important is analysis of the barriers impeding the best use of this information to adapt. The Climate Commission, an independent body set up in early 2011 to provide reliable and authoritative information on climate change, and to inform the debate on this issue of national significance, is young in its responsibilities. It would be of great value if it evolved as a trusted channel of communication from the scientific community to the general public. It would also be of great value if it evolved into a source of information for government. While there is a substantial body of research on climate change mitigation to aid policy makers, there has been relatively little research on adaptation. This limits the ability to identify 'no-regret' measures that would be justified under all possible future climate change scenarios. And it affects our ability to identify measures that reduce our vulnerability to climate change while meeting other policy objectives.

At the moment it is difficult for government to answer questions about how well we are adapting. And although climate change risk is gradually being reflected in government approaches in non-climate policy areas, we are not able to say whether we are adapting enough and in the right ways. Consequently, it is difficult for governments to evaluate where best to direct its efforts to reduce barriers hindering efficient adaptation. Also, it is difficult to assess whether the cumulative result of decisions is a better adapted Australia.

Adaptation policy and the regulatory role of government in infrastructure

Some of the necessary regulatory roles of government intersect with adaptation to climate change. The government as owner of some types of infrastructure, as regulator of others and with responsibility for land-use planning will necessarily be at the centre of many adaptation decisions.

The Australian Government's assessment in 2009 of climate change risks to Australia's coast provided for the first time a nationwide indication of the extent of risk, with up to $63 billion of existing residential buildings alone at risk from inundation by 2100. Further work is needed to identify risks to essential services and infrastructure, and to the commercial sector. However, there is clearly a large legacy risk in the coastal zone. Eventually, the impacts on the coast could lead to abandonment of houses, resettlement of towns or the construction of major protection works for threatened cities and public facilities such as airports.

The recommendations from the National Climate Change Forum held in February 2010, followed by the report to the Australian Government by the Coasts and Climate Change Council in December 2010, highlight the need for national action to help coastal communities (including those outside capital or major cities) prepare for the impacts of climate change, as there are significant economic and social implications of adaptation.

The report notes that, without coordinated action, there is an increased chance of inefficient and wrongly focused adaptation—of actions that, while delivering short-term benefits, may exacerbate vulnerability to climate change over the longer term. The forum concluded that national action was needed to enhance consistency in policy and regulatory settings across jurisdictions, and identified a number of key issues—sea-level rise planning benchmarks, risk guidance for planning and development, legacy issues and legal liability, building codes and standards, and integrated regional planning approaches. A major barrier to adaptation identified by the forum was moral hazard—the expectation that government will support those whose property is damaged by an extreme event—which presents a disincentive to prepare for future risk.

The government in 2009 identified initial national adaptation priorities—coastal management, water, infrastructure, natural systems of national significance, disaster resilience and agriculture. The Council of Australian Governments' agreement in February 2011 on the National Strategy for Disaster Resilience demonstrates an increased focus on emergency planning and the implications of climate change for disaster preparedness and highlights the change in emphasis from reactive responses to proactive risk-reduction measures.

The Commonwealth, state and local governments are responding with increasing awareness and forethought to the climate change adaptation challenge, despite the raucous public disputation over whether climate change is a problem that warrants attention at all. It is good to know that Australians have not lost our characteristic ability to respond pragmatically to real problems when we see them, undisturbed by disputation over dogma. But we are in the early stages of thinking through all of the implications for government of effective adaptation to climate change.

Biodiversity and ecosystems

Climate change is a significant and additional pressure on ecosystems and biodiversity in Australia. It will affect ecosystems and biodiversity by shifting, reducing and eliminating natural habitats. In Australia, many species of flora and fauna are at risk from rapid climate change because of their restricted geographic and climatic range. Where ecosystems and species have low tolerance for change, altered climatic conditions can trigger irreversible outcomes such as species extinction.

Just as greenhouse gas emissions without a carbon price represent a market failure, the decline in Australia's biodiversity can be attributed at least in part to a failure to correct through public policy the market's failure to value the natural estate. This failure, combined with the vulnerability of Australian ecosystems to climate change, provides a strong argument for the establishment of market mechanisms to ensure the resilience of Australia's ecosystems. For example, the Henry tax review pointed to the important role government can have in protecting biodiversity and ecosystems through specified payments, for example, in management agreements with landholders.

There is increasing private philanthropic interest in maintaining biodiversity, but government is likely to remain the major source of funds to conserve biodiversity. Separate but complementary incentives for carbon sequestration and other ecosystem services will allow the respective benefits to be sold in separate markets, with landowners selling into both and making decisions that maximise total incomes and benefits to themselves.

Conclusion

The 2008 Review discussed a number of challenges for Australia in a changing climate in the areas of water scarcity, risks to infrastructure, resilience of ecosystems and biodiversity, and disaster resilience. Developments since 2008 continue to highlight the importance of these issues, including the need for further reform to reduce barriers to adaptation. The type and extent of adaptation will be affected by the characteristics of the climate risk, the decision makers and the institutional framework within which adaptation decisions are made.

We need to think in a more coherent and integrated way about how we allocate inevitably large sums to adaptation. The inclination will be to respond to each crisis separately. And yet the increased challenges of extreme events of flood and fire and drought, of disruption of infrastructure in heatwaves, of erosion of coastal properties, of changes in fish stocks and disappointment about sustained river flows for irrigation, are different aspects of a single phenomenon.

An integrated adaptation response with clear priorities will be of particular importance where there are long-lived decisions to be made on land-use planning and major infrastructure development.

Australia's future economic productivity will be influenced by the ability of the government to provide climate change information and develop tools that can be used at the appropriate scales for decision making by private agents, and to develop coherent approaches to land-use planning and to management and climate-relevant building codes and other standards in high climate risk areas.

Australians in future will have to manage the world as they find it. We may be leaving them with a difficult task. We should seek to avoid leaving them with an impossible one.

We will improve their chances by encouraging an effective global mitigation effort and doing our fair share; reducing emissions in the lowest-cost way through carbon pricing; replacing the great Australian complacency of the 21st century with a new era of productivity-oriented reform; working to establish and extend effective markets generally and in insurance, water and food in particular; strengthening applied climate change science and making its results widely available; being cognisant of the value of our own inheritance of biodiversity and reflecting that value in our decisions on managing climate change; and embodying knowledge of climate change in private and public infrastructure decisions.

PART III
AUSTRALIAN TRANSFORMATIONS

9 Innovation nation

DR ZHENGRONG SHI, chief executive officer of Suntech, the world's largest solar photovoltaic company, recently wrote an article titled 'Can Australia save the world?'. Dr Shi observed that the United Kingdom had shown the world the way to use coal for energy and that the United States had shown the world how to harness atomic power. He asked whether it would be Australia and China that would show the world how to best use solar power.

The answer to Dr Shi's question is a resounding 'yes'.

Dr Shi is an Australian citizen and former researcher at the University of New South Wales' School of Photovoltaic and Renewable Energy Engineering. The school has had a quite extraordinary impact on the global photovoltaic industry. Four of the top six global manufacturers in solar photovoltaic technology are linked with the University of New South Wales. Beyond Suntech, there is the world's second largest manufacturer, JA Solar, also founded by former researchers at the school. Trina Solar, the fourth largest, was founded by one of its PhD graduates. The technology of the sixth largest, Yingli Green Energy Holding, was piloted by another graduate of the school.

In fact, the influence is so great that the school refers to these graduates and former researchers as its 'gigawatt club'—the group of former staff and students whose global firms now produce more than a gigawatt of solar products a year.

The story of the School of Photovoltaic and Renewable Energy Engineering has a moral: it shows how the economic benefits of 'spillover' in innovation can accrue to all nations—when one country creates a breakthrough in technology all others stand to benefit.

The transition to a low-carbon economy will be a story of innovation. The costs of the transition will depend on how effective we are in discovering and applying new technologies for producing goods and services with fewer emissions; or in satisfying demand in ways that produce fewer emissions, or which sequester carbon dioxide and store it safely.

There are multiple motives for innovation in some of these areas. Increased efficiency in the use of energy, the development of new energy sources for electricity and the accumulation of carbon in soils may all lead to lower costs independently of the need to reduce emissions. The rising costs of oil—with depletion of the limited stock of natural resources that are readily and cheaply accessible—strengthen this motive for energy saving and alternative energy. This motive has been further strengthened recently by the large increases in current expectations of future oil prices.

Some firms and countries will undertake or encourage innovation in these areas because they see themselves—firm or country—as producers of goods embodying the technology, like Suntech. For some countries and firms, there may be anxiety that others will gain earlier access to superior new technology and receive competitive advantages in the marketplace.

Countries that are large importers of fossil fuels and face future price increases may see development of emissions-saving technologies as a way of reducing prices. When the imports are especially important to domestic economic stability, as they are with oil, and are drawn from places that carry political risks, the motives for innovation that reduces reliance on imports may include national security.

And now for governments, and for firms as well if the costs of carbon emissions are subject to a price, there is the motive of reducing greenhouse gas emissions.

All of these motives were woven into the main theme of President Barack Obama's State of the Union address in January 2011:

> Meanwhile nations like India and China ... are investing in research and new technologies. Just recently, China became the home to the world's largest private solar research facility and the world's fastest computer ...

> This is our generation's Sputnik moment. Two years ago, I said that we needed to reach a level of research and development we haven't seen since the height of the Space Age ... We'll invest in, especially, clean energy technology—an investment that will strengthen our security, protect our planet, and create countless new jobs for our people ...

> Already we're seeing the promise of renewable energy ... We're telling America's scientists and engineers that if they assemble teams of the best minds in their fields, and focus on the hardest problems in clean energy, we'll fund the Apollo projects of our time ...

> So tonight, I challenge you to join me in setting a new goal: By 2035, 80 percent of America's electricity will come from clean energy sources ...

> So instead of subsidising yesterday's energy, let's invest in tomorrow's.

It is unlikely that the other motives alone will go anywhere near reducing greenhouse gas emissions enough, and in a short enough timeframe, to avoid great damage from climate change. Moreover, there would be no reason for any investment at all in technologies to store carbon dioxide wastes in geological structures if we provided no incentives for emissions reductions and relied on other motives.

Placing a price on emissions of greenhouse gases that reflects the damage that they do to other human activities is the economically efficient way to increase incentives for innovation in technologies that reduce greenhouse gas emissions. It will increase the expected profitability of all such activities, increase the levels of innovation, and speed it up. No useful area of innovation to reduce emissions will miss out on the encouragement. It will add to other motives for investing in innovation, and lead to higher levels of investment in innovation than the other motives alone are encouraging.

President Obama was not reacting to shadows in referring to international competition for low-emissions technologies in his State of the Union address. There is heightened awareness that the leading industrial countries are engaged in a great race to find the technologies that will carry the world to a low-carbon economy. There is awareness that future generations will be using energy from different sources and in different ways than we do. Firms and countries will need to produce goods and services that make sense when those different ways have become business as usual.

I have conducted recent correspondence with a senior researcher at the Development Research Centre of China's State Council (Cabinet), Yongsheng Zhang, on China's interest in innovation in low-emissions energy. The Development Research Centre and the World Bank are jointly conducting a research project on green energy in future Chinese development. There is acute awareness in China that the long-term development path of China will be greatly affected by its success in innovation to reduce the costs of low-emissions energy. In addition to facilitating implementation and then extension of China's own international commitments on emissions, it will determine the role that China can play as a supplier of capital goods to a low-carbon global economy.

In the remainder of this chapter, I look at how Australia fits into the global innovation story, and discuss policies for innovation in Australia.

Pinpointing market failures

The carbon price will make it more profitable for firms and industries to invest in research, development, demonstration and commercialisation of low-emissions technologies. It guides and provides incentives for investments in low-emissions technologies.

It is impossible to know in advance where investment in innovation will occur or whether it will be successful. Entrepreneurs will form their own views and back them with investment in the full awareness that they are taking risks. The leaders of public entities that provide fiscal support for innovation will also be making decisions under great uncertainty. The advantage of a broad-based market instrument like a carbon price is that it will draw out the most prospective low-emissions innovation across the Australian economy. In much the same way that such a mechanism identifies least-cost abatement, a carbon price is the most efficient stimulus for innovation.

But these positive effects alone will not be enough to generate economically desirable levels of investment in innovation.

When a private firm invests in research, development, demonstration or commercialisation of new technologies, it takes large risks and spends money on discovering knowledge. If it is successful, it reduces risks and discovers knowledge from which it will receive some benefits in future, but which other firms will share. Patents can keep a proportion of the benefits within the innovating firm, but sometimes only a small proportion, and only for a while. The benefits that one firm's innovation confers on others justifies public subsidy—without public support, there will be much less innovation than is desirable from the point of view of the community as a whole.

Innovation is especially valuable at a time of large and rapid changes in relative prices and in economic structure. In these circumstances, private expenditure on innovation falls short of socially valuable levels by an especially large amount, so the case for public subsidy is especially strong.

To take advantage of the new opportunities provided by a carbon price and to reduce emissions at low cost, substantial public support for innovation is required. Economically valuable innovation has national and international dimensions. This is clear from the Suntech case. The benefits of investment in research, development, demonstration and commercialisation of new technologies are not generally confined within national boundaries. Australian firms will eventually benefit from successful innovation in, say, new biofuels technology that is developed elsewhere. But other Australian firms are likely to benefit more quickly and perhaps more comprehensively from innovation that is undertaken successfully in Australia.

What follows from this international character of the external benefits from innovation?

One consequence is that there may be too little public support for innovation if it is left to the isolated decisions of individual countries—just as there is likely to be too little investment in innovation if it is left to private entities alone without public fiscal support. Sovereign governments will provide support for innovation on the grounds that there will be substantial 'spillover' benefits within their own territories. Indeed, the national advantage from one country establishing itself as a major global centre for production of goods and services embodying a new technology may be large enough to encourage a high level of activity. But we are more likely to obtain a globally optimal level of investment in innovation if each national government is confident that others are making large contributions.

Since the Great Crash of 2008, just such a shift has taken place, with many countries turning to substantial 'green' stimulus spending. This has reversed the 35-year decline in real terms in low-emissions energy research, development and demonstration. Stimulus spending saw such investment by governments of developed countries grow from US$15 billion in 2008 (in 2009 prices) to US$23 billion in 2009. The major contributors were the United States, at around US$12 billion, and the European Union, at around US$6 billion.

But, just as in other dimensions of the mitigation project, overall research and development spending in the major developing countries is growing more rapidly still. Chapter 4 discussed China's use of stimulus spending in response to the Great Crash to accelerate development of low-emissions technologies. The growth in the general level of China's official research and development spending has continued more or less in line with its high economic growth rate. This growth in investment easily outstrips rates in all other countries, and is expected to continue. The Indian Government has recently established a National Clean Energy Fund for research and innovation, which is financed from production and imports and is expected to provide at least US$550 million per year.

While the increase in government financial support should drive innovation, the International Energy Agency has cautioned that the global impetus for investment in this area through the 2008 and 2009 stimulus packages may not be sustained as governments of the developed countries seek to restore order to national budgets, citing a lack of major announcements in the first half of 2010. Indications for 2010 show that spending levels have dropped and were closer to 2008 levels, marking the end of stimulus spending.

There are good reasons for high-income countries to play their proportionate part in a global innovation effort. That part will be most productive if each country contributes in areas in which it has a comparative advantage in research.

Developed countries have superior endowments of relevant human and physical capital for successful research and development. They are also in a better position than developing countries to invest in long-term and risky projects that hold out the possibility of high returns.

As leading American economist Jagdish Bhagwati has argued influentially in the Indian discussion, developing countries could see a commitment by developed countries to low-emissions innovation and dissemination as a way of discharging their historical responsibility for exhausting the planet's capacity to absorb greenhouse gases. This adds to the case for developed countries to accept commitments to provide fiscal support for research, development and commercialisation of new technologies, whether at home or in developing countries.

In the 2008 Review I noted the requirement for investment in low-emissions technologies of around $US100 billion per year to reduce the costs of transition to a low-carbon economy. I proposed a Low-Emissions Technology Commitment, within which developed countries would undertake to provide their share of a global effort on this scale, calibrated according to national income. Australia's share was calculated at $2.8 billion, which comes down to about $2.5 billion with the exchange rates and other parameters of 2011. Each country would be free to allocate these funds according to its own priorities for work to be undertaken at home or in developing countries.

Australia should commit to its share of the Low-Emissions Technology Commitment, building up to $2.5 billion per year over several years. This funding would be allocated to a number of measures to support increased investment in the research, development, demonstration and commercialisation of low-emissions technologies. Measures to lower the cost to Australia of the transformation to a low-emissions economy should include increasing support for public and private basic research, market-led support for private demonstration and commercialisation, and strong and independent governance arrangements.

There are already a number of government expenditures in these areas, which are harder than might be imagined to separate out in the budget allocations. My recommendations on innovation are based on the expectation that current commitments to expenditure on low-emissions technology innovation during the forward estimates period should be maintained, and that the presumption be made that such expenditures—to the extent of three-quarters of a billion dollars per year—would have been continued beyond the forward estimates throughout the ten-year carbon revenue budget period, without drawing on carbon revenues.

Existing arrangements include the Australian Government's $5.1 billion Clean Energy Initiative. The capacity to make good use of innovation expenditure would rise over time so budgetary provision is phased up towards $2.5 billion per year. Carbon revenues would be used to fund the gap between expenditure required under the Low-Emissions Technology Commmitment and expenditure within the established budgetary arrangements—the gap above three-quarters of a billion dollars in the later years.

Over the first five years of the carbon pricing arrangements, funding committed under these programs would meet a substantial part of the fiscal support for innovation. The funding from the general revenue would be gradually brought within the governance arrangements for innovation support from the carbon revenue. Funding from general revenue could be redeployed in the light of experience, subject to contractual and other indelible commitments. In later years the majority of the innovation support would be funded from carbon price revenue.

New or modified governance arrangements are required to ensure that funds for innovation in low-emissions technologies are used effectively. While a number of steps have been taken towards stronger and more independent governance arrangements, the recommendation from the 2008 Review to establish a new low-emissions innovation council remains relevant today. The council would have oversight of programs across all areas of innovation relevant to mitigation. Ultimately, it would be preferable to have a single overarching body to administer programs for all technologies that will play a role in lowering Australia's emissions.

After a time, the case for accelerated investment in innovation in this particular area would decline, as the market caught up with the sudden expansion of opportunities created by the introduction of the carbon price. Australian fiscal support for innovation in low-emissions technologies could then ease back towards general levels of innovation support, a decade or so after the introduction of carbon pricing.

After a period of adjustment to the carbon price—a transitional period of perhaps ten years—the special case for higher funding for innovation in low-emissions technologies will probably have run its course. However, this should be assessed closer to the time. Beyond this transitional period, funding for innovation in low-emissions technologies can be made through the economy-wide measures available generally to support research and development.

For the rationale for exceptionally large fiscal support for firms that invest in new low-emissions technologies to be sound, government must be able to assure the Australian community that its approach to innovation support is efficient, effective and likely to yield a net benefit to society.

At the basic research end of innovation, there is no alternative to governments, and independent experts on behalf of governments, making decisions on the projects to which public funds will be allocated. Market forces cannot drive Australia's public organisations towards the most beneficial projects in basic research. Government will obtain the best results if it entrusts the task of selecting projects to receive government research funds to a well-equipped independent body. The goal of the body will be to allocate finite resources towards areas of research that will generate large national benefits if they are successful and where Australia already has a strong capability. Comparative advantage in research and national interest in deployment should be the main criteria for allocating funding for basic research.

At the demonstration and commercialisation stage, government can rely on market processes to pick those projects that have the best chance of success and are likely to generate large gains if successful, and are therefore most worthy of taxpayer support. Good governance and sound criteria are of central importance to this approach.

There are three strong reasons for supporting basic research and development. First and foremost, basic knowledge is a public good: once new basic knowledge is created, it is impossible for the person or firm that created it to contain the value or capture all the benefits.

Second, a range of other benefits arise from basic research, principally in the ongoing development of the labour force through concurrent education and training. Third, basic research often entails collaboration, which in turn generates benefits that exceed the sum of the individual research parts. Sometimes this collaboration extends across disciplines and institutions and the parts are only combined in the most productive ways through institutional change. Also, building basic research capacity enables faster resolution of intractable problems that typically arise when developing complex first-of-a-kind technology systems; solving these problems often requires a basic research breakthrough.

The economic case for investment in basic research and development is uncontroversial and widely accepted. The Productivity Commission opposes research and development support purely for the sake of fostering infant industries for good reasons. But it accepts that where underinvestment is a bigger problem in an emerging industry than an established one, more government support could potentially lead to better outcomes for society.

At the demonstration and commercialisation stage, the primary market failure is spillovers—the costs faced by early movers who make the initial investment to demonstrate or apply new technologies that benefit the industry more widely. They can include the costs associated with training in new skills; working through new regulatory frameworks; development of supporting industries and a reliable supply chain; demonstrating and communicating the safety and effectiveness of new technologies to the community; and educating providers of debt and equity about the technical and commercial dimensions of a new technology.

Public funding of education and training can help overcome these barriers, while regulatory and legal barriers to innovation can be reduced with foresight and active policy.

Learning by doing

As a new industry or sector develops and expands, it uncovers cost reductions and more efficient approaches to technology deployment. For example, on-the-ground learning in Spanish solar thermal manufacturing and deployment has led to a fourfold increase in the speed of parabolic mirror assembly, significantly lowering the cost of the product overall.

Costs can also be expected to fall because of economies of scale. As global demand for new technologies increases, production will also shift from batch engineering to more efficient mass-production processes. The cost reductions from mass manufacture of solar photovoltaic panels is a well-documented example of this phenomenon—it is estimated that there has been a twentyfold increase in manufacturing capacity in China in just four years. Unit costs have been falling rapidly with the increase in scale. The cost of nuclear power plants in China has fallen well beyond earlier expectations as multiple orders have allowed production of components and construction to be placed on a continuous basis rather than being produced to order.

Cost reductions can also accrue when multiple identical projects occur in the same geographic area—also known as the local convoy effect—potentially delivering a 5 to 15 per cent reduction in capital costs. Discussion with industry suggests that this is a significant driver of higher costs for wind generation in Australia, where wind farms tend to be dispersed across rural areas, according to the availability of spare transmission capacity.

Learning by research can create a step change in technology cost curves. Significant cost reductions are possible in the shift from the use of parabolic mirrors to concentrating solar thermal towers to produce solar thermal energy more efficiently by achieving higher operating temperatures. Some analysts suggest that the rate of technological advance in concentrating solar thermal could make it competitive with conventional generation sources in the next five to ten years.

Technology costs

The resources boom has lifted Australian costs in general, which has affected the cost of new as well as established technologies. The prices of labour, goods and capital that go into generating new technologies are all higher than expected in the 2008 Review. This contrasts with generally falling capital costs for low-emissions technologies in most of the world.

The rapid growth in new energy technologies has placed a strain on supply and raised the costs of a number of raw materials. For instance, the rising price of polysilicon raised the cost of photovoltaic modules. The requirement for rare earth materials for the production of batteries is another example. This will generally be a short-term effect, as higher prices will provide incentives for large increases in supply and subsequently for lower prices.

The cost pressures have been offset in this country to a degree by the strong Australian dollar, which has reduced the costs of many of the imported components of low-emissions technologies.

Nonetheless, the general market constraints in Australia for materials, skills and finance have created a temporarily high level of price inflation on top of the 'real' technology cost curves. Within this context, however, several low-emissions technologies in the electricity and transport sectors have the potential for surprising rates of cost reduction as a result of innovation.

The remainder of this chapter contains four examples of areas of technological innovation that have particular interest to Australia, given the nature of our economy and natural endowments. The first two, carbon capture and storage and biofuels and biosequestration, are areas in which Australia has a strong comparative advantage in research as well as a strong national interest in application. The third, solar energy, is an area where Australia has some strengths in research capacity through which it has made, and will continue to make, globally significant discoveries, but in which it will generally be a user of the successful outcomes of overseas research, development and commercialisation. Finally, there is a brief update on developments in technology in the transport sector. Here Australia will generally be applying technologies developed elsewhere, although again, we will make contributions to technological innovation in niche areas.

The application of technologies that have been developed elsewhere in a new setting always requires innovation. The maintenance of some world-class research capacity in a field is generally helpful to early adoption of technological innovation from abroad.

Carbon capture and storage

In 2008, it seemed that carbon capture and storage technologies would be a viable and substantial part of the suite of future low-emissions technologies. Its eventual emergence on a large scale was built into the Garnaut–Treasury modelling of the costs of mitigation. Studies and trials to date indicate that there are no insurmountable technological challenges.

Carbon capture and storage has been applied in commercial contexts for several decades and carbon dioxide has been injected in geological reservoirs to enhance oil recovery; several large-scale sequestration projects are building on this experience.

The Gorgon Carbon Dioxide Injection Project, currently under construction in Western Australia, is an important example of capture and sequestration in the process of gas liquefaction. On completion, it will be the world's largest geosequestration project. The project will cost approximately $2 billion and is an integral component of the $43 billion Gorgon Liquefied Natural Gas Project. It will inject between 3.4 and 4 million tonnes per year of carbon dioxide equivalent into a geological formation which is more than two kilometres underground. This would account for nearly 1 per cent of Australia's annual emissions. This is about one-quarter of the emissions of a large brown coal-fired power station (for example, Hazelwood in Victoria), so carbon capture and storage is already important in Australia.

The successful expansion of carbon capture and storage in gas would be of considerable significance. Fugitive emissions from gas and coal currently account for around 7 per cent of Australia's emissions and are expected to account for around 25 per cent of total emissions growth to 2020 under current policies. The contribution would be expanded considerably if geosequestration of emissions from gas combustion were added. Even if carbon capture and storage were applied to emissions associated with gas liquefaction alone, it could make a substantial contribution to Australia's mitigation effort.

The deployment of carbon capture and storage in the electricity generation sector, especially where coal is the energy source, is more technically challenging and expensive. There are large differences in capture costs depending on underlying costs of energy and the distance from and quality of storage sites. The cost of energy matters because capture and storage require large amounts of energy. The Latrobe Valley has significant advantages, with an especially cheap energy source and proximity to an excellent and well-known geological structure, the Gippsland Basin. But other areas with less propitious circumstances face high costs. The effects of high costs have been exacerbated by the economic challenges relating to climate policy uncertainty and first-of-a-kind technology risks.

In recent years, several prominent demonstration projects in the electricity generation sector have failed—including the $4.3 billion, 400 megawatt ZeroGen project in Central Queensland and the US$2.2 billion, 275 megawatt original FutureGen project in the United States. For the ZeroGen project, the primary hurdle was the difficulty of locating an appropriate geological formation. Exploration for bankable storage sites to serve large-scale demonstration projects can be at least as costly and risky as oil and gas exploration. The second large challenge is in accurately estimating the costs of large-scale projects.

Despite initial disappointments in the electricity sector, this is not an unexpected path for the development of such a challenging and complex technology. The G8 goal of broad deployment of carbon capture and storage by 2020 remains achievable, but will be challenging and require political leadership at all levels of government. Governments have made commitments to support around 25 large-scale projects worldwide with a significant increase in allocated funding in 2010. In total, governments have now committed up to US$40 billion to support carbon capture and storage demonstration projects. And the funding allocated to specific large-scale projects is expected to double in the next couple of years.

But we may be close to the point where the risk of disillusionment will accumulate to prohibitive levels in relation to geosequestration from coal-based electricity. This would seriously diminish the prospects for effective action against climate change. It would also seriously diminish the viability of coal as a long-term energy source in Australia's export markets as well as in Australia. It would be unfortunate if a technology with such potential for Australian and global mitigation and such significant implications for the future of the coal industry were abandoned before it had been tried in favourable locations and conditions.

Biofuels and biosequestration

Australia has been at the forefront of research and development in the biological sciences related to the land sector for more than a century. Australian rural industries are built on continuous application of the results of research and development in a country in which soils and climatic conditions are more challenging for agriculture and forestry than in most of the rest of the world.

Australia also has immense opportunities for the absorption of carbon into soils, pastures, woodlands and forests. These are discussed at length in Chapter 10. Here it is enough to observe that research, development, demonstration and commercialisation of new technologies in the land sector should be a major focus of research efforts and public expenditure related to the transition to a low-emissions economy. Australia has a major role to play in the global system in relation to research on measurement of carbon in land environments, biosequestration technologies and practices, and the development of domestic and international rules for incorporating reductions in emissions associated with rural activities into national and international systems of incentives for mitigation.

Australia also has an important role to play in research and development on biofuels.

Biofuels using traditional agricultural land as a source of biological inputs are problematic, largely because they displace food crops. But the new (second-generation) biofuel production systems, which use advanced technologies and non-food plant materials, do not have these problems. They offer the potential for significant emissions reductions compared to fossil fuels and some existing (first-generation) biofuel production systems. The feedstocks for these new systems include algae, crop and forestry residues, and purpose-grown non-food agricultural products. Biofuel products include ethanol, butanol and biodiesel.

These new sources of biomass can be produced on less productive land, allowing relatively low production costs, avoidance of competition with food production, and new commercial opportunities for landholders.

Algae is particularly promising, because of its high efficiency in converting the sun's energy and carbon dioxide into hydrocarbons. It requires much less land or sea area than plants to convert a specified amount of carbon dioxide. Biofuel production technologies that use algae as feedstocks will allow Australia to use resources that it has in abundance—sunlight and saline water and land. There is significant research being undertaken on algae as a feedstock within public research organisations, and commercialisation efforts within several private firms. These warrant strong support.

The new biofuel production systems are at varying stages of development, with numerous pilot projects in operation around the world. Good progress is being made internationally on research, development and commercialisation for some technologies—large-scale production of ethanol from lignocellulosic material is predicted to become cost-competitive (without subsidies) with fossil fuels by 2015.

Australia's relatively modest investment in biofuel research, development and commercialisation deals with new production technologies and transport infrastructure suited to Australia's environment, land management systems and transport fuel needs. Further investment in innovation is needed across the production chain, from biomass availability and harvesting through to processing and fuel production technologies, vehicle performance and distribution infrastructure.

Solar energy

There are two well-established solar energy technologies: solar thermal (including concentrating solar power) for large-scale power generation; and solar photovoltaic. In Australia, solar thermal water heating has been the predominant form of solar energy use to date, with solar photovoltaic representing only 5.8 per cent of total solar energy consumption.

The global photovoltaic market has exploded in the last decade, with an average annual growth rate of 40 per cent. Significant cost reductions have been associated with the increase in installed capacity, linked to both technological improvements and economies of scale. A considerable proportion of the total cost of installing a solar photovoltaic system is represented by the array of photovoltaic cells known as the 'module'. Photovoltaic modules have displayed a well-documented historic learning rate of 22 per cent almost consistently from 1976 to 2010, while capital costs have fallen by 22 per cent for each doubling of capacity. No other energy technology has shown such a high rate of cost reduction over such a long period. Costs are continuing to fall at a rapid rate with expansion of large-scale production in China.

Reflecting the global trend, Australia's total photovoltaic peak generation capacity has increased fivefold over the last decade, driven partly by support through the Solar Homes and Communities and Remote Renewable Power Generation programs. Domestic prices of photovoltaic systems have dropped as a result of an increase in international competition among a larger number of suppliers (influenced by rapid growth of solar energy in China), increased scale of production and a strong Australian dollar. There are still considerable cost-reduction opportunities for photovoltaic systems in both technology improvements and efficiencies of scale, with capital costs expected to fall by 40 per cent by 2015 and 70 per cent by 2030.

The form of solar thermal known as 'concentrating solar power'— a less mature technology than photovoltaic—has also made considerable progress in recent years. A range of sources agree that it has significant cost reduction potential, based on known technical improvements, economies of scale and the increase in industry knowledge from continued deployment of the technology, similar to the observed learning rates of solar photovoltaic. Forecasts of capital costs in the short term are similar to the expectations set out in the 2008 Review, but the longer-term projected cost reductions are significant.

While the vast majority (96 per cent) of concentrating solar power plants built to date have been parabolic troughs, analysis by the CSIRO and others shows that power towers (with a central receiver) have the potential to achieve the lower cost.

Concentrating solar power has been deployed globally since the 1980s and is now undergoing a resurgence, particularly in Spain and California. Its inherent advantages include high efficiency of energy conversion; easy integration with low-cost thermal storage to provide renewable power well into the evening demand peak; and use in conjunction with fossil fuels (notably gas) using the same boilers and generators. Such hybrid generation increases the steadiness of output and reduces the cost of power, and provides more uniform output compared to other intermittent renewable technologies.

The advantages of concentrating solar power and the prospects for rapid cost reductions in both solar photovoltaic and concentrating solar power hold promise for the global mitigation challenge.

Fast trains and electric cars

The 2008 Review discussed the many ways in which emissions would be reduced in the transport sector in the transition to a low-carbon economy. Innovation would be important, but Australia would be mainly absorbing technologies from abroad. (The main exception is probably the role that Australia could play in the development of new feedstocks for biofuels, discussed above.)

The continuation of high oil prices and increasing road congestion in our large cities have helped to sustain the growth in demand for public transport that was evident three years ago. The use of public transport in all states is constrained by capacity. The shift to more emissions-efficient smaller cars with a leavening of hybrids has continued.

The most important overseas developments in the cost of new transport technologies relate to mass production of fast trains in China, and the development of electric vehicles and reduction in their costs everywhere at a much faster rate than had been anticipated in 2008.

China used its stimulus package in response to the Great Crash of 2008 to bring forward from 2020 to 2012 its plans for constructing 13,000 kilometres of fast train track. This has brought into existence a formidable supply capacity that will in due course reduce the costs of deploying fast trains outside China, including in Australia.

There have been major developments relating to electric cars. Zero-emissions road vehicles now seem set to be the most promising source of abatement in the transport sector, through the interaction of electrification of vehicles with the decarbonisation of electricity. The 2008 Review projected that electric vehicles would account for 14 per cent of the transport task in Australia in 2050. Since then there have been many signs that the penetration of electric vehicles will proceed much more rapidly than was built into the Garnaut–Treasury models.

There are a number of reasons for this accelerated transition.

First, stimulation of demand for electric vehicles in the European Union, United States and China has turned out to be stronger than expected. A range of subsidies, tax credits and other incentives have created this effect.

More importantly, after a period of intense debate between the automakers and governments, stricter vehicle emissions regulations are becoming the norm. China's fuel economy standards rank third globally behind Japanese and European standards. President Barack Obama, in his 2011 State of the Union address, announced an objective of having one million electric vehicles on American roads by 2015. Targets announced by major economies would see the sale of more than 20 million electric vehicles by 2020.

Third, direct multibillion-dollar government investments in battery and electric vehicle research and development are leading to a faster rate of technological development and associated cost reductions. Supply-side factors are also playing a role in accelerating electric vehicle penetration. Economies of scale, design improvements and technological improvements are driving battery production costs down. Some analysts have noted that battery costs have been declining more rapidly than expected. Through its multibillion-dollar investment in batteries and electric vehicles, the Obama administration is projecting cost reductions of more than 80 per cent by 2020 along with significant concurrent improvements in battery performance and durability. For a number of years before the global financial crisis, Toyota made batteries the main focus of its large research and development effort in anticipation of pressures for reduction in emissions, and its early lead still puts it in a strong position for hybrid vehicles.

After several years of progress by private Chinese investors seeking to jump ahead of rivals in a new automotive technology where they have no disadvantage from being a latecomer, they have recently been the recipients of major venture capital investment from the United States. In May 2011, Volkswagen announced that it was building a plant for electric vehicles in Shanghai.

The economics of petrol versus electricity as a source of fuel have changed with higher oil prices. In the 2008 Review, forward oil prices in the US$60–70 per barrel range were built into the modelling, but global forecasts now tend to be in the range of US$120–130 per barrel. Strong growth in developing country demand and limited opportunities for expanding supply raise the possibility of prices going much higher at times, pending the development of alternative sources of energy for transport.

One central determinant of future rates of adoption of electric vehicles will be the ability to finance and roll out extensive charging networks. The many hundreds of millions of dollars already committed by private investors suggest that finance will not be a barrier. Australian investors in this area emphasise the potential for the storage capacity of batteries in electric vehicles to transform management of power demand costs and so to lower the costs of generation and distribution of power.

Transport emissions are likely to be reduced earlier and at a lower carbon price than had been anticipated, as long as there is early decarbonisation of the electricity sector.

Conclusion

Past experience with market-based approaches to pollution control in Australia and overseas suggests that government forecasts tend to underestimate the rate of commercial innovation and thereby overestimate the costs of such schemes to society once adequate incentives for innovation are in place. Recently, the Chinese authorities have been surprised by the rate at which the costs of some low-emissions technologies—wind, solar and nuclear—have fallen. Industry projections from overseas suggest that there is the potential for these costs to fall substantially in the short to medium term.

The falling costs of new low-emissions technologies generally bode well for the global and Australian transitions to low-emissions economies.

Technological change can substantially reduce the costs of these transitions to a low-emissions economy. We cannot anticipate the shape or the extent of that change before it unfolds, as firms and individuals find new ways to respond to incentives to economise on emissions. We can, however, put in place the policies that will encourage individual investment in emissions-reducing innovation.

The central policy instrument to encourage the use of established low-emissions technologies and to discover and to apply new technologies is carbon pricing. Putting a price on carbon increases the profitability of investment in innovation.

But the carbon price alone will not lead to adequate investment in research, development and commercialisation of new technologies, because the private investor can capture only part of the benefits. Fiscal incentives can bridge the gap between benefits to the whole of society and benefits to the individual investor in innovation. Part of the carbon pricing revenues—on the plateau of expenditure between about five and ten years from the commencement of carbon pricing, about $2.5 billion per year of the Australian revenue—can be used productively for this purpose.

Support for innovation should extend from basic research and development to the demonstration and commercialisation of new technologies. The basic research will be conducted mainly but not only through public institutions. It requires decisions on allocations of expenditure according to assessments of Australia's comparative advantage in research capabilities, and national interest in successful outcomes. At the commercialisation end of innovation, allocations are best guided by private priorities backed by private commitments of funds, in the form of matching grants or other benefits from government.

10 Transforming the land sector

GLOBAL FOOD security has been a hot topic since the large spike in world food prices in 2008. Between 2006 and 2008, global food prices rose by 60 per cent. The spike is estimated to have increased by 100 million the number of people considered to be 'food insecure'. Demonstrations and riots occurred in more than 30 countries across the Middle East, Africa, Asia and Latin America, and in Haiti food riots led to the toppling of the prime minister.

Global food prices eased following the Great Crash of 2008. The easing was brief. Prices again surged in 2010 and have risen to new heights in the first months of 2011. They were one factor behind political unrest in the Middle East in the Australian summer of 2010–11.

One long-term source of upward pressure on food prices has been strong growth in demand for high-quality food with economic growth in China and other successful developing countries, recently reinforced by climatic disruption in China and South Asia. The high food prices over the past year have been driven also by an unusual range of other severe climatic events affecting global agriculture: dry conditions in the United States; floods in Australia, Canada, Pakistan and Brazil; dry conditions in Argentina; and high temperatures, drought and wildfires in Russia. Once world grain prices started to rise strongly, the increase was exacerbated by a number of countries, most importantly Russia in the recent episode, seeking to enhance their own food security by restricting exports of grain.

The less successful the reduction of emissions over the years ahead, the more climate change is expected to disrupt global food production. But poorly designed mitigation can also generate large food problems. Significant increases in the production of biofuels driven by government mandates, said to be motivated by climate change mitigation, particularly in the United States and Europe, have contributed to food price increases through recent years. In the United States around 40 per cent of the corn crop and in Europe nearly 40 per cent of canola is now used to produce biofuels. Demand for biofuels is estimated to have accounted for 60 per cent of the global change in demand for wheat and coarse grains between 2005 and 2007. According to a recent study, setting a global biofuels target of 10 per cent of transport fuel would lead to the number of people at risk of hunger rising by 15 per cent, while only delivering significant emissions benefits after 30 to 50 years. When the use of biofuels is mandated, there can be large losses of food production with minor or no reductions in total emissions—an outcome that would be avoided with economy-wide carbon pricing.

The recent succession of extreme weather events and the associated large losses of food production have focused attention on how future climate change is likely to affect world food production and food security. A major study by the International Food Policy Research Institute estimates that, with the effects of climate change and 'middle of the road' estimates of future incomes and population, prices for wheat, rice and corn in 2050 would be high even with perfect mitigation and much higher without mitigation.

Recent food history

Relative to the prices of other commodities, food prices fell through most of the 20th century. There had been pessimism in the early 1960s about whether the acceleration in population growth that characterised the early years of the second half of the century would overwhelm growth in food production. This was one spur to greatly increased public investment in agricultural research in the large developing countries, which joined traditionally high investments in research in the developed countries. The research yielded what is now known as the 'green revolution', a series of technology developments and transfers around the world that steeply increased global agricultural production.

As a result, food supplies easily outran the rapidly growing population in the last third of the 20th century. The number of people living in extreme poverty (living on less than US$1 a day) fell steadily, and encouraged the development of the United Nations' ambitious millennium development goals, which include accelerated reduction in the number of the world's poor and hungry people. Most important of all, the spreading of modern economic development into the large developing countries and the commitment of the Chinese Government to radical fertility-control policies from early in the reform period in 1980 saw a rapid reduction in fertility rates and population growth. The last of these developments promised permanent victory of food supplies over population.

However, the Platinum Age has taken us into a new, different and bigger race between human requirements for food and its supply. The pressure of modern economic activity on the earth's natural resources and environment is growing far more rapidly and on a much larger scale in the early 21st century than at any time in history. At the same time, rapid economic growth and higher living standards are doing three things that hold out prospects for humanity's success. The Platinum Age is reinforcing the continued reduction in fertility and population growth. It is widening access to the knowledge and technology that can break the link between growth in living standards and pressure on natural resources and the environment. And finally, it is lifting much of humanity rapidly towards the conditions in which concern for environmental amenity rises in priority alongside consumption of goods and services.

This new race is between humanity's increased pressure on the natural environment on the one hand, and its capacity and will to break the nexus between high and rising material standards of living and pressure on the natural environment on the other. An optimistic view of this new form of an old race suggests that it will be a close-run thing.

Climate change is just one of the points of vulnerability of the natural environment in the 21st century. But it seems at this stage to be the one that has by far the greatest potential to trip up humanity in the great race.

The higher food prices are triggered by climate change mainly through lower yields. According to the study by the International Food Policy Research Institute, with climate change, wheat yields in developed countries are 4.2 per cent lower in 2050 and 14.3 per cent lower in 2080, and in developing countries are 4.1 per cent lower in 2050 and 18.6 per cent lower in 2080. Incidentally, declines in production are disproportionately large in Australia, deriving from lower yields and contraction of planted areas.

In the study's optimistic case, climate change increases the number of malnourished children in the world by 10 per cent above what it would otherwise have been in 2050. In its pessimistic case, there are 9 per cent more malnourished children (from a higher base).

The study concludes that good policy and innovative production responses could significantly alleviate the pressures of climate change. But the effects of unmitigated climate change from 2050 to 2080 would possibly be unmanageable. The most important elements in an effective adaptive response would be the maintenance of good policy for broad-based sustainable development; large increases in investment—including public investment in agricultural research—to raise productivity; global free trade in food; and decisive early action on adaptation to and mitigation of climate change. The other items on this list do not look much easier to accomplish than policies to hold climate change within moderate bounds.

Australian consequences

People with good incomes can afford to meet their food needs even when food prices are high, and whether or not their own country produces food. People in Singapore and Hong Kong probably have as much or more fundamental food security than any in the world. They buy large quantities of the best quality food from almost all of the substantial food-producing countries.

A sound economy generating income security provides the first pillar of food security.

Australia is rich enough for all Australians to have incomes that are adequate to purchase their essential needs for good food if policy is set with that objective in mind. It is difficult now to imagine circumstances in which Australians would face shortages of food.

Difficult, that is, until we work through the implications of changes in temperature and rainfall and its variability as a result of climate change, and imagine other countries closing their food markets and banning exports as food prices rise under the influence of difficult climatic conditions. We were given a foretaste of that possibility with the bans on grain exports in many countries during the price spike of 2008, and with the Russian bans on wheat exports in the 2010 drought and wildfires.

An open, rules-based international trading economy, built around maintaining free trade in agricultural as well as other products, is the second pillar of food security. A sound economy and a sound international trading system based on principles of free trade are usually enough to guarantee food security.

The real prices received by Australian farmers declined substantially in the second half of the 20th century, while the real prices paid for inputs remained roughly constant. This resulted in a substantial decline in Australian farmers' terms of trade and a substantial fall in the real net value of farm production in the 1980s and 1990s.

The high global food prices of 2008 and 2010 affected Australians but were not a threat to Australian food security. The high agricultural prices of recent times—which can be expected to rise further in future even with effective climate change mitigation, and further still in its absence—have been a boon to some Australian farmers.

The real value of Australian farm production rose again in the decade to 2011, following favourable conditions in the early 2000s. However, later in the decade, it resumed its downward movement under the influence of prolonged drought.

It may not seem remarkable that Australian farmers' terms of trade have been roughly maintained through the last decade. But the change in trend is of truly historic significance. The huge appreciation of the Australian dollar during the resources boom has crushed the international competitiveness of the services and manufacturing sectors, and in old circumstances would have had a similar effect on the farming sector. This effect would have been exacerbated if the long-term fall in global agricultural prices that reduced Australian farmers' terms of trade by more than half between 1950 and 2000 had continued into the early 21st century.

As the modelling by the International Food Policy Research Institute shows, in the absence of climate change, global food prices (and with them, Australian farmers' terms of trade) could have been expected to rise over the next half century.

With climate change, however, the outlook is more complex, and problematic. World agricultural production will be affected adversely by warming, by more variable weather, and by more intense severe weather events. That will raise global food prices. The price increases in themselves would help Australian incomes. But Australian food production would be affected more than food production in most other countries, and most Australian farmers would be battling to maintain the production that would allow them to take advantage of higher prices.

The 21st century is likely to provide large opportunities for Australian farmers if there is effective global mitigation. It is likely to be deeply problematic without effective global mitigation. The Australian rural economy has an immense interest in the success of global mitigation.

Farmers are naturally anxious about the costs that they will bear as a consequence of Australia doing our fair share in strong and effective global mitigation. Like all Australians, and more than most, they will contribute through higher costs of fuel and transport. Like all Australians, they will be beneficiaries of adjustments to tax and social security that emerge from the allocation of revenue from carbon pricing. The rural community, like all Australians, will be intensely interested in the details of the allocation of increased costs and tax cuts associated with the introduction of carbon pricing.

Australian farmers have no interest in the failure of global mitigation—this would deny many of them the opportunity profitably to expand production and exports in a period of rising global prices.

We will probably have to live through an extended period of transition to global carbon pricing. An initial period, in which farmers can take advantage of the sale of genuine abatement whether or not it complies with Kyoto rules, would provide potentially large opportunities for augmenting rural incomes.

We will only learn how large by introducing the opportunity and observing what use is made of it.

So the world of the 21st century, with Australia playing its part in an effective global mitigation effort—and providing incentives for land-related sequestration at the general carbon price—would be a good world for Australian farming.

Many Australian farmers managed to stay afloat and some to prosper during the long decline in agricultural terms of trade in the second half of the 20th century. Those who remain in farming today have revealed a capacity for innovation based on advanced knowledge, sound management and flexibility in the face of changing opportunity. This same quality will provide the basis for great success in times of expanded opportunity.

If there is no effective global mitigation, success will require a lot of good luck at the farm level. At the national level, Australia would depend on imports for food from time to time, and perhaps much of the time. We would depend on an open global trading system not only for export income, as we do now, but for food security. We may have no choice but to make the most of such a world. But both food security and farmers' income security make it a choice to avoid if we can.

Farming Cancun

Greenhouse gas emissions from the land sector in Australia are affected by patterns of agriculture and forestry as well as by traditional management by Indigenous Australians and nature conservation. The land sector is a source of several greenhouse gases, including carbon dioxide, methane and nitrous oxide. It also provides biosequestration, which is the removal of atmospheric carbon dioxide through biological processes—photosynthesis in plants—and storing carbon over significant periods of time.

The land sector of the economy is a significant part of Australia's emissions profile, because of our large land area and agricultural and forestry resources relative to population. The agriculture, forestry and other land use sectors contributed around 18 per cent of total annual emissions in Australia's most recent Kyoto Protocol accounts. Globally, the land sector accounts for around 26 per cent of emissions. It is the main source of emissions in many developing countries.

The rules for the land sector under the Kyoto Protocol provide limited incentives for countries to reduce emissions and increase biosequestration.

Australia has been one of the countries seeking to broaden the coverage of biosequestration within the international rules. Our objective should be comprehensive carbon accounting, with sound principles for managing emissions due to natural events beyond a country's control. The development of sound principles in domestic coverage of land-based sequestration will be influential in the international discussions.

The negotiations in Cancun in 2010 made progress on some of these issues. If there is international agreement on improvements to the current rules, it could boost the national economic value of Australia's land-based mitigation options.

Australia is exposed to greater risks of damage from climate change than any other developed country. Because of our large land mass and small population, we also have greater opportunities for low-cost mitigation in the land sector than any other country. So we should do what we can to encourage more comprehensive accounting for emissions within the international system. We could advance our interests by adopting more comprehensive accounting ourselves at an early date. This would provide a model for the productive evolution of a global system that provides for appropriately expanded coverage of emissions from the land sector.

Easing into a carbon price

The government has introduced into the parliament legislation to establish the Carbon Farming Initiative, an offset program targeted to begin from July 2011. Under the government's proposal for a carbon pricing mechanism released in February 2011, emissions from sources covered under the Carbon Farming Initiative would be excluded from coverage under the carbon pricing mechanism. Credits for reductions in emissions that count towards Australia's Kyoto Protocol target could be used in the carbon pricing mechanism, or alternative funding arrangements could be adopted for the land sector.

Resolution of accounting rules and estimation issues will open up greater opportunities for emissions reductions and biosequestration in the land sector. Eventual movement toward international acceptance of the full range of genuine land-based sequestration and full coverage of the land sector under a carbon pricing mechanism could substantially reduce the cost and encourage the raising of mitigation ambition for Australia. In the meantime, however, Australian farmers could be disadvantaged if they were forced to adopt practices that made it more expensive or difficult for them to compete internationally. This is the general problem of the trade-exposed industries, but one that is particularly challenging in this sector.

The problem is more acute in this sector than in manufacturing or mining, where other countries are already applying substantial and costly constraints on emissions. Most individual farm businesses generate emissions on too small a scale for them to be covered by the general carbon pricing system proposed by the 2009 Carbon Pollution Reduction Scheme. Measurement is costly and difficult on a small scale in the present state of knowledge. To date only New Zealand has committed to including agriculture in an emissions trading scheme by 2015. That date is subject to review. New Zealand's commitment is significant for Australian farmers, as New Zealand is Australia's main competitor in domestic and the most important international markets for meat, dairy products, wool and temperate horticultural products.

The 2008 Review recommended that the land sector initially be brought within incentives to reduce emissions through offsets, and brought within an emissions trading scheme once issues regarding emissions measurement, estimation and administration are resolved. The proposed date for inclusion of New Zealand agriculture (2015) is a good time for a review of whether circumstances have changed enough for Australia to have full coverage of the land sector.

The government's proposed Carbon Farming Initiative is an important first step in encouraging abatement in the rural sector. It will provide valuable lessons in Australia and internationally on the administration of land sector incentives. It will also lead to 'learning by doing' improvements in technologies applied to emissions reduction and sequestration in the land sector. The government's proposed design provides some encouragement of new emissions reduction and biosequestration, while constraining the risks of giving credit for activities that do not deliver real abatement.

Under the Carbon Farming Initiative, landholders will be able to submit projects for approval on a voluntary basis. They will be able to sell offset credits from a range of approved activities, so long as legal obligations such as periodic reporting are met. The new scheme covers emissions reductions and biosequestration in agriculture, forestry, other land uses and landfill waste deposited before 1 July 2011. This coverage is likely to break new ground for offset schemes not only in Australia, but internationally. Of all established offset schemes, only the Alberta Offsets Scheme in Canada covers a similarly broad range of emissions offsets from agriculture.

Under the Carbon Farming Initiative, it is proposed that offset credits from biosequestration and emissions reductions will be based on the net emissions or removals each year as measured against a baseline. The baseline represents the emissions that would have occurred in the absence of the incentive provided by the initiative. Originally this proposal included requirements to establish that financial considerations would not have been adequate to encourage sequestration in the absence of the value generated by the scheme. The Review's fourth update paper suggested suggested the exclusion of this consideration, and the government's scheme has been amended to accept the advice. It is important that the setting of baselines should not disadvantage early adopters. The government has acknowledged this issue in consultation papers on the Carbon Farming Initiative.

The design proposed for the Carbon Farming Initiative applies a common framework for crediting abatement that would count towards Australia's Kyoto Protocol target, and non-Kyoto abatement. This approach would provide for broad coverage and, depending on the opportunities to obtain a significant price for abatement, could encourage greater mitigation than an approach that applied different rules to Kyoto and non-Kyoto abatement. It avoids landholders having to interpret Kyoto Protocol rules and provide evidence of, for example, areas of vegetation that met Kyoto forest criteria in 1990. This is a sound approach.

Permanence is a critical issue. Unlike other emissions reductions, the abatement achieved through biosequestration can be reversed by events that are natural as well as by human action. While permanence can never be guaranteed, the risks of an unplanned release of emissions can be reduced through good system design and good management practices. In addition, new insurance products are emerging that are applicable to carbon forests. These may offer forest growers risk management options that complement the management of risks to permanence proposed within the Carbon Farming Initiative.

The government proposes a rigorous but flexible approach to dealing with permanence. Participants wishing to change land use must relinquish credits. A proportion of credits from biosequestration projects would be withheld as a form of insurance against losses. Where unintended losses occur, credits would not have to be relinquished as long as the project was re-established. New credits would not be issued until the previous levels of accumulated sequestration had been reached.

I suggested in March 2011 that genuine land sector abatement should be rewarded with a credit that was equal in value to the carbon price within the formal scheme. For activities that are currently recognised for Australia under the international rules, credits could be sold to parties that were liable under the scheme. For other genuine biosequestration, the regulator of the carbon pricing scheme would purchase credits at the carbon price. Sales of land-based credits under these arrangements would be placed under generous quantitative limits. Sales of credits that complied with current international rules would be limited to 4 per cent of permits in the first year of the scheme, rising to 10 per cent in 2020. Credits that fell outside current international rules would be limited to 2 per cent of the value of scheme permits in the first year, rising to 4 per cent in 2020. These quantitative limits would be removed when there was full coverage of land sector emissions under the scheme.

These limits would allow for high values of sales of credits from the land sector. If the limits were fully utilised, the total value of sales of credits in 2020 would be in the vicinity of $2.25 billion, or roughly the value of Australian wool production in the most recent full year.

The land and farm management practices that would generate credits under the recommended arrangements could have substantial additional benefits to farm productivity and the environment. The build-up of carbon in soils is helpful to retention of moisture and therefore for the maintenance of farm production in times of drought. It assists productivity by improving the availability of nutrients needed for plant growth. Restoration of woodlands can have benefits for stock management and, if accompanied by appropriate incentives of other kinds, for biodiversity.

The land sector could make a large contribution to the reduction in Australian emissions. Analysis by CSIRO indicates that technical potential (the upper limit of what could physically be possible) is about twice the total level of current Australian emissions. The realisation of a small percentage of that potential would make a significant difference. Let's take a look at the opportunities.

Deforestation

Deforestation is undertaken for agricultural purposes, as well as for mining, urban development and infrastructure such as roads and powerlines.

Since 1990, there have been large reductions in deforestation rates in Australia, and therefore emissions, due to economic, technological and climatic factors as well as government regulation. In regions such as semi-arid Queensland, where there has been extensive clearing for livestock grazing since the middle of last century, regrowth of woody vegetation following clearing is common. Landholders clear regrowth once it has reached an extent that reduces livestock production. Since 1990, the amount of clearing of regrowth has increased relative to the area of first-time clearing.

Further abatement could be achieved by reducing the rate of deforestation and retaining or promoting regrowth vegetation on land that has already been deforested.

The greatest opportunities for reducing deforestation and for maintaining and promoting regrowth are in Queensland, and, to a lesser extent, in New South Wales.

CSIRO's analyses indicate that a carbon price or an offset credit at around $15–25 per tonne of carbon dioxide equivalent would encourage landholders to retain substantial areas of native vegetation that might otherwise be cleared. Landholders would make decisions based on carbon price levels, possible impacts on agricultural production and ecosystems, and risks of loss of stored carbon (for example, due to fire or drought). In some instances, decisions will also be influenced by the need to adapt to climate change.

Livestock

When cattle and sheep digest their food, they produce methane emissions, which account for about 10 per cent of total national emissions. These emissions have declined by 14 per cent since 1990, largely because of a fall in sheep numbers due to the combined effects of the extensive drought and a fall in the price of wool relative to other agricultural products.

Several abatement options are available. These include changes in animal breeding, diet and management.

Improving emissions estimation methods to allow assessment of the impacts of emissions reduction options will be essential for further developing and realising abatement strategies for cattle and sheep and other livestock.

Soil

Compared to northern hemisphere soils, many Australian soils have naturally low soil carbon levels due to their old, weathered nature and the effects of a warm and dry climate. Large losses of soil carbon have occurred since the conversion of native vegetation to agricultural land started in the 1800s.

Over recent decades, Australian farmers have progressively adopted practices that reduce soil disturbance and reduce losses of soil carbon, such as no-till and conservation farming practices. These practices also reduce production costs and land degradation in cropping systems. Adoption levels for these practices have reached 90 per cent in many regions, and there have been rapid increases in the last five to ten years in some regions where adoption had previously been relatively low.

As well as reducing losses, the amount of carbon in soils can be increased, for example through establishing pastures using perennial species and adding organic matter such as manure and green wastes.

There are risks that increases in soil carbon can be reversed, for example in drought. Drought caused a significant spike in national emissions from croplands during 2002 and 2003.

There is also considerable interest in the potential for incorporating biochar into soils to increase soil carbon. Biochar can be produced through a combustion process at high temperatures from sources including wood, agricultural crop residues and green waste; gas produced in its creation can be used to generate electricity or converted to liquid fuels. Biochar has greater stability than the material from which it is made, and can therefore provide a long-term carbon store. It can be added to soils, and may improve soil fertility, which could in turn provide biosequestration benefits through enhanced plant growth.

Recent studies have confirmed earlier indications that some types of biochar can significantly increase crop yields, and some are stable in soil for decades, although these qualities vary with the feedstocks and production processes used. Crop yield responses to addition of biochar are also variable, ranging from a 30 per cent reduction to a 200 per cent increase, and can vary with soil type. The introduction of financial incentives for increases in soil carbon allows farmers to take up the opportunity in those places where it is profitable for them to do so.

CSIRO estimated that building soil carbon, combined with nitrous oxide emissions reductions, on cropped land had a national technical abatement potential of 25 million tonnes of carbon dioxide equivalent per year from 2010 to 2050. This assessment assumed adoption of practices to improve crop productivity and reduce tillage across 20 million hectares of annually cropped soils.

Rangelands

Arid and semi-arid rangelands, which include grasslands, shrublands and woodlands, make up about 70 per cent of Australia's land mass, or about 550 million hectares. Rainfall in these rangeland areas is highly variable.

Over many years, marginal sheep and cattle grazing has caused considerable degradation of some of these rangelands, including shrublands and woodlands dominated by mulga.

The most likely way in which rangelands will be rehabilitated is through reducing grazing intensity. Other possible ways to rehabilitate rangelands and increase carbon levels include introducing or re-establishing palatable shrubs such as saltbush, tagasaste or other perennial shrubs, and fire management.

Recent studies have indicated substantial, but widely differing, technical abatement potential from rehabilitating degraded rangelands. Most studies apply similar sequestration rates of less than one tonne of carbon dioxide equivalent per hectare per year through rangeland rehabilitation. The differences in technical potential arise primarily from differences in the land area considered by each study. Some of the differences in area derive mainly from differences in definition.

CSIRO estimated that rehabilitating 200 million hectares of overgrazed rangelands could have a technical potential to sequester 100 million tonnes per year of carbon dioxide equivalent between 2010 and 2050.

Fire management in northern Australia

Tropical savannas cover the northern third of Australia and are largely owned and managed by Indigenous Australians. They include grasslands and woodlands and are used by Indigenous peoples for traditional purposes, and for grazing and conservation. Fires are common in savannas, especially in the late dry season, when fuel loads are highest and have dried out. Strategic burning earlier in the dry season can help reduce fuel loads so that late dry season fires are smaller and less intense. Intense, hot fires late in the season burn more completely, and damage trees and native fauna in ways that controlled early season burning does not.

Savanna burning is the major source of emissions in the Northern Territory. Australia's total emissions from savanna burning have declined since 2001. Emissions are heavily influenced by climate factors from year to year.

Improved management of savanna fires has been estimated by CSIRO to have the technical potential to reduce emissions by 13 million tonnes of carbon dioxide equivalent per year, or around 90 per cent from the average level over recent years, between 2010 and 2050.

Another assessment of abatement options on Indigenous land estimated that strategic fire management had the potential to reduce emissions by 2.6 million tonnes per year. The West Arnhem Land Fire Abatement Project is operating successfully across 28,000 square kilometres. The cost of implementing the project has been estimated at between $7 and $30 per tonne of carbon dioxide equivalent. Commercial viability will vary across regions and depend on the carbon price.

Plantation forests

Australia's plantation forest estate has expanded significantly since 1990. An average of about 64,000 hectares per year was established from 2002 to 2008. Within this average, there was a decline from 72,000 hectares in 2008 to 50,000 hectares in 2009.

CSIRO assessed the technical abatement potential of plantations established since 1990 to be 400 million tonnes of carbon dioxide equivalent per year between 2010 and 2050, with a carbon price of $20 per tonne and an average carbon sequestration rate of 9 tonnes per hectare per year.

CSIRO cautions that this estimate needs to be considered as an upper limit once market demand, processing capacity and transport costs are taken into account. Other constraints on expansion include willingness by landholders to convert agricultural land to forest, regulatory restrictions on forest establishment, the transaction costs of carbon market participation, and impacts of climate change on land productivity for forestry.

Some of these constraints could be addressed by growing native tree species on low productivity land for carbon sequestration as well as harvesting for timber or biomass energy production. Research suggests that there are some prospects for growing low rainfall plantation eucalypt species.

Native forests

The area of forests and wooded lands per person in Australia greatly exceeds that of other developed countries. Native forests cover around 147 million hectares, or almost 20 per cent of Australia, including 23 million hectares held in conservation reserves and 9.4 million hectares of public land where timber production is permitted. The remaining area comprises public land used for other purposes, and privately owned land. The area of native forest harvested has declined over time, and totalled about 81,000 hectares in 2009.

There is limited information on carbon sequestration in native forests, and current estimates are subject to significant uncertainties. Taking these uncertainties into account, CSIRO has estimated that if native forest harvesting were to cease, there is a technical potential for abatement of 47 million tonnes of carbon dioxide equivalent each year from 2010 to 2050.

There has been considerable recent discussion among forest industry and union representatives, environmentally focused non-government organisations and governments on greatly reducing harvesting of trees in native forests on government-owned land. The sequestration benefits of such developments are considerable. Where arrangements are negotiated for reduction of native forest harvesting and a financial settlement is made, future claims on the carbon scheme revenue should be excluded.

More comprehensive carbon accounting could open up opportunities for carbon markets to provide a source of revenue for forest managers. Emissions reductions and biosequestration in harvested native forests could be achieved by reducing the area harvested, or potentially through changes in harvesting practice. Forests that are subject to minimal human influence are likely to be either mature or regrowing following fire or other natural disturbance, and therefore provide limited opportunity for active management to increase carbon storage.

Carbon forests

Carbon forest plantings are grown for the purpose of biosequestration, and are a relatively new activity. They include plantings of mixed native species as well as single species such as mallees, and are often designed to provide other benefits for biodiversity, natural resource management and farm productivity. Plantings may be established in blocks, widely spaced rows or in ways designed to provide specific environmental benefits, for example along stream banks or as corridors for native species. Australian companies that are managing new forest plantings to provide emissions offsets commonly use locally native species.

As carbon prices rise, establishing forests in regions with lower rainfall and lower land values becomes economically viable. Carbon plantings are more likely to be suited to these growing conditions than timber plantations. CSIRO has estimated that, with a carbon price of $20 per tonne carbon dioxide equivalent and incentives for biodiversity benefits, establishing biodiverse carbon forests could have a technical biosequestration potential of 350 million tonnes of carbon dioxide equivalent per year between 2010 and 2050. At a large scale of activity, some carbon forest establishment could replace growing of forests for wood production.

In recent years at least 20 businesses and not-for-profit organisations have been reported as offering carbon forest offsets in Australia. New agreements between carbon forest growers and companies with large emissions profiles for extensive plantings to offset energy emissions indicate expanding capacity and readiness in the carbon forest industry ahead of a carbon price.

Bioenergy

Using bioenergy (liquid biofuel, electricity and heat) instead of energy derived from fossil fuels can deliver mitigation benefits if emissions over the lifecycle of production of the biomass feedstock and energy are lower than for fossil fuels. The biomass feedstocks can be sourced from purpose-grown agricultural and forestry crops as well as waste material. Biofuel can also be produced through gasification of biomass. Biofuel production supplies less than 1 per cent of Australia's total transport fuels.

Biomass feedstocks for heat and electricity in Australia currently comprise by-products or residues from agriculture and forestry production systems. The major feedstocks for heat and electricity are sugar cane waste and wood and wood waste. Biomass energy contributes around 1 per cent of Australia's total electricity generation.

Biomass electricity is financially competitive with other renewable technologies where the fuel costs can be kept relatively low. Unlike some other prospective low-emissions technologies, current industry estimates do not envision significant potential for capital cost reductions or significant increases in the scale of operation for power generation using biomass only. There are opportunities in the transition to a low-carbon economy for biomass to be used with coal in power generation to lower emissions.

The use of planted mallee eucalypts as a biomass source for energy and other products has been investigated for a number of years, and has been demonstrated in a pilot bioenergy plant in Western Australia. Unlike some agricultural sources of biofuels, the ratio of energy output in biomass to energy inputs in production is highly positive. Commercial viability of growing mallee for bioenergy would be enhanced by innovation to reduce growing, harvesting and transport costs. Cost reductions of 50 per cent are expected within ten years.

Combining sequestration with biodiversity

There is increasing recognition in Australia of the value of biodiverse forests and woodlands. Incentives for carbon sequestration will incidentally encourage biodiverse development in some circumstances, and work against it in others. While establishing or restoring a native forest or woodland might support a rich and diverse ecosystem, the mass planting of a single species of tree would not.

If carbon pricing is combined with separate incentives that recognise the value of biodiversity, market decisions will generate combinations of sequestration and biodiversity that have maximum value. The ideal arrangement is for separate funds to be established for rewarding investments in biodiversity, and for landowners to be able to draw both on these and carbon pricing arrangements. Established state and federal schemes to encourage biodiversity could form the basis of the required arrangements.

Land and water

Reducing emissions and increasing biosequestration in rural Australia will involve some changes in land and water management. Many changes will be positive: improved soils, restored habitats and new sources of regional income. There could be negative impacts, not just on biodiversity but also on water resources and agricultural land use.

In our market economy, landholders should be free to use their land as they judge best for themselves unless there are good reasons for the community as a whole to constrain private choices. Carbon pricing will cause farmers to substitute some sequestration activities for conventional farming simply because it is profitable to do so. On the other hand, rising food prices could favour food production over forests. In many cases the changes will be new mixes of activity rather than complete changes in land use; farmers in the Western Australian wheat belt have been planting rows of mallee trees since the 1990s to reduce land degradation while maintaining crop production.

There are roles for regulation, as well as market-based approaches, in helping to avoid unintended negative impacts. Land-use planning requirements already apply to plantation forestry. Decisions on whether to constrain changes in land use are best made at a local level, and should aim to achieve a balance between land uses that is appropriate in each local community.

Conclusion

The land sector is greatly affected by climate change and has a large part to play in its mitigation. This is true for the world as a whole. It is more powerfully true in Australia than in any other developed country.

The world has entered a challenging period of rising food prices in the 21st century, after a long period of decline. This presents problems for global food security. These challenges can be met, so long as the higher food prices are not compounded by the effects of weakly mitigated or unmitigated climate change.

In themselves, higher food prices represent opportunities for the Australian rural sector.

The Australian rural sector is set to do well in these new circumstances if the world is effective in mitigating climate change. The measures that reduce emissions in the rural sector will add considerably to rural incomes once they are rewarded within carbon pricing arrangements. Some will confer substantial benefits for farm productivity and for adaptation of farm management to more intense drought and other extreme weather conditions that are associated with climate change.

In the absence of effective global mitigation—in which Australia will have to do its fair share—the 21st century will be deeply problematic for global and even Australian food security, and for the income security of rural Australia.

The international rules developed within the Kyoto process overlook many potentially important areas of land sector mitigation. The omissions are especially important in Australia. Australia has a major role to play in developing alternative and economically and environmentally more efficient rules governing incentives for mitigation in the land sector. Demonstration of their suitability in Australia can lead to their adoption in other countries, including our developing country neighbours with their large forestry sectors.

The land sector, especially through biosequestration, has immense technical potential for reduction and absorption of emissions. Realising a small proportion of that potential through providing incentives commensurate with the sector's mitigation contribution would transform the Australian mitigation effort. It would also greatly expand the economic prospects of rural Australia. Complementary incentives for biodiversity would help to ensure that the potential for carbon and biodiversity efforts to assist each other is realised.

We are a long way from knowing how much of the technical potential can be realised economically. The linking of the proposed Carbon Farming Initiative with the carbon pricing scheme would open the way to realisation of that potential. This is an essential modification of the proposed Carbon Farming Initiative. Once it was linked to the carbon pricing scheme, the Carbon Farming Initiative would reveal the potential and define the extent to which it is economically relevant; it does this by providing for the emergence of an offset market for land sector abatement.

In time, as the world shifts towards pricing carbon in farming, the Carbon Farming Initiative can be merged with the broader carbon price and fulfil its full mitigation potential.

11 Electricity transformation

IN 1865, William Stanley Jevons, a founder of modern economics as well as meteorology, published a book called *The coal question*. In it he drew attention to the United Kingdom's limited coal supplies and commented that 'if we lavishly and boldly push forward in the creation and distribution of our riches, it is hard to overestimate the pitch of beneficial influence to which we may attain in the present. But the maintenance of such a position is physically impossible'.

Jevons did some of his seminal work in Australia and his words have a special resonance for contemporary Australians. The largest source of Australia's disproportionately high greenhouse gas emissions is our reliance upon coal in the electricity sector. As such, we face the same choice outlined by Jevons. We might elect to do nothing and continue to enrich ourselves, in part with cheap energy. But in the long run climate change ensures that the maintenance of our current prosperity under business as usual remains impossible.

The transformation of Australia's electricity sector is thus about ending reliance on fossil fuel long before the coal runs out—unless we can capture and safely store the carbon dioxide wastes.

Australia's unusually emissions-intensive electricity sector is the main reason why Australia's emissions per person are exceptionally large. The transformation of the electricity sector has to be at the centre of Australia's transition to a low-emissions economy for this reason, and also because the lowest-cost path to reducing emissions in the transport, industrial and household sectors involves greater use of low-emissions electricity.

A carbon price that passes through to household and business electricity prices will drive the reduction of emissions in the electricity sector. This will involve a switch in the predominant sources of power used in generation. It will involve the building of new low-emissions generation and the closure of high-emissions generation. It will also moderately reduce the growth in electricity demand in the short term, and more strongly over the longer term as people economise on electricity use.

For households, carbon pricing will raise the price of electricity. The price increases associated with the introduction of a carbon price come at a difficult time. There have been large recent electricity price rises that are not related to a carbon price, and without changes in the regulatory arrangements this would continue. The increases are mainly because of large investments in the networks of poles and wires that distribute electricity, and the high rates of return on those investments that are recouped without risk from consumers. These investments have been stimulated by a regulatory regime that provides excessive incentives for investment whether or not it is wanted by consumers. The effects of the resources boom on coal and gas prices and construction costs are likely to increase electricity costs and prices further still.

This circumstance can be corrected. There are strong signs that lower growth in demand is reducing the need for investment. And the inefficiencies in domestic energy markets and regulatory regimes that underpin the rises can be corrected to ease the adjustment to carbon-related price rises. Indeed, These things should be done anyway. The introduction of carbon pricing has drawn attention to their importance.

Price rise drivers

Australian households and businesses enjoyed relatively stable and low retail electricity prices for many decades. After a long period in which Australian electricity prices rose more or less in line with other prices, from 2007 to 2010 prices rose nationally by 32 per cent in real terms. While the consumption of electricity makes up a relatively small component of a typical household's expenditure, rises of this magnitude put pressure on lower-income households.

Electricity prices for businesses have also increased rapidly since 2007. Household and business electricity prices have not always moved together, but recent price rises are common to both.

Electricity price increases are set to continue under current policies and regulatory arrangements. In those states in which electricity prices are regulated for residential and household consumers, further price increases have been announced. New South Wales and Tasmania have made decisions on electricity prices beyond 2011. In New South Wales, annual electricity prices are expected to rise by around 18 per cent in 2011–12 and by around 10 per cent in 2012–13. Electricity prices in Tasmania are expected to increase by around 10 per cent annually over 2011–12 and 2012–13. These expectations do not contain allowance for a carbon price.

The increases in electricity prices reflect many factors including increased investment in electricity networks—the poles and wires that distribute electricity from power plants to the home—as well as policy changes (such as the Renewable Energy Target) that have led to higher costs.

One way to explore the increase in prices is to examine how costs have changed for the three components of electricity prices—the costs of generating the power (turbines), the cost of distributing it to households (poles and wires) and the cost of retailing (marketing the product).

The costs of generating power accounted for around 40 per cent of the overall electricity price in 2010 (see Figure 11.1). The cost of moving that power to households—transmission and distribution—made up about 50 per cent of the price. The energy retailers accounted for 10 per cent of the price.

In the current period it is distribution and transmission costs that are the greatest factor in rising electricity prices, accounting for approximately 68 per cent of recent price rises. Retail costs are also increasing much more than the average.

Figure 11.1: Electricity costs and their contribution to current price rises in 2010

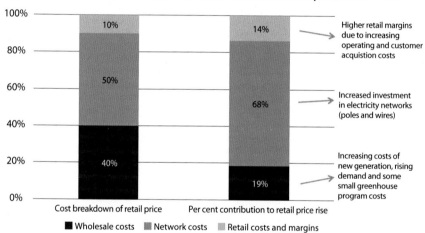

Note: The contribution of the cost components to electricity price rises is based on an average of the current regulated retail price determinations across jurisdictions in the National Electricity Market (except in Victoria, which no longer regulates prices).

Sources: Essential Services Commission of South Australia 2010, *Review of retail electricity standing contract price path—final report*; Independent Competition and Regulatory Commission 2010, *Retail prices for non-contestable electricity customers 2010–2012*; Independent Pricing and Regulatory Tribunal of New South Wales 2010, *Review of regulated retail tariffs and charges for electricity 2010–2013*; Office of the Tasmanian Economic Regulator 2010, *Final report—2010 investigation of maximum prices for declared retail services on mainland Tasmania*; Queensland Competition Authority 2010, *Final decision—benchmark retail cost index for electricity: 2010–11*.

In most of Australia, the generator market is competitive and therefore wholesale prices are determined primarily by the dynamics of supply and demand. As consumer prices have risen in the past three years, there has been an easing in the growth in demand. As well, over the past year, milder weather reduced summer demand and industry sources also suggest that the insulation program and photovoltaic installations have had some effect. There have been price fluctuations, in part because of drought as the costs of water-cooled coal-fired power stations rose and because of a reduction in output of the Snowy and Tasmanian hydro-electric systems. The end of the drought placed downward pressure on generator prices from mid-2010.

Electricity network costs, on the other hand, have marched higher on the back of a surge in investment. Electricity networks are split into the transmission network and the distribution network. Transmission is the extremely high-voltage assets—metal towers connecting generators to substations. Distribution is the lower-voltage wiring that brings power from substations to customers. Both are regulated under similar rules.

Transmission network investment over the current five-year regulatory period is forecast at over $7 billion and $32 billion for distribution networks. This represents a rise in investment from the high levels of the previous period, of 84 per cent and 54 per cent (in real terms) in transmission and distribution networks respectively.

These high levels of network investment have been attributed to the need to replace ageing assets, electricity load growth and rising demand, as well as rising peak demand and changed standards in reliability and service requirements.

This explanation raises questions. Demand growth has been slow in recent times, long before the cooler summer of 2010–11 (see Figure 11.2). Why does old investment from the 1950s and 1960s suddenly have to be increased now? Certainly there has been growth in peak demand, but this is avoidable: other countries provide high incentives to reduce energy demand at the peaks, while Australian regulatory settings reward distributors for growth in peak demand.

Figure 11.2: National energy demand

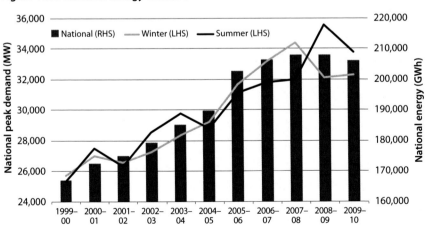

Source: Australian Energy Regulator 2010, *State of the Energy Market 2010* and www.aer.gov.au/content/index.phtml/tag/MarketSnapshotLongTermAnalysis.

A second explanation for the rising network costs is that several states have recently adopted higher reliability standards for distribution networks. These require additional capital investment by the network businesses in these states to ensure that the higher standards can be achieved within the regulatory requirements.

The setting of reliability standards and service requirements has not been subject to institutional or regulatory reform. We already have a reliable system. It is important that disciplines are introduced that balance consumers' interest in low prices with marginal improvements in reliability.

This marginal increase in reliability comes at a cost that is paid by all electricity consumers, and not necessarily valued at anything like their cost by many of them. There is no opportunity for consumers to make their own choices on what they are prepared to pay for greater reliability, when standards are already high.

Price rises have also been stimulated by other government policies. Measures to promote energy efficiency improvements and renewable energy generation are funded from the prices paid by consumers for electricity. These policies therefore contribute directly to higher retail electricity prices, and depending on the level of uptake, have the potential to place further upward pressure on prices. They feed into all three components—wholesale, network and retail—of electricity costs.

The first of these other government policies is the Renewable Energy Target scheme, in which retailers must ensure that a proportion of their electricity supply comes from renewable energy sources. Renewable energy is currently a more expensive source of electricity and therefore adds to wholesale electricity prices. Unlike economy-wide carbon pricing, the Renewable Energy Target does not necessarily encourage the lowest-cost means of reducing emissions. Nor does it encourage innovation: it favours the lowest-cost established technologies that are eligible within the scheme.

Another policy is feed-in-tariff schemes, which pay a premium rate to encourage renewable electricity generation through small-scale generation such as solar photovoltaic systems. The costs of these premium rates are spread across all consumers.

Benefits of a national network

There is one more source of electricity price rises that must be addressed if we are to maximise the impacts of the carbon price and the fight against climate change. Currently, there are distinct electricity markets in each region of the

National Electricity Market. The interconnectors between them frequently constrain interstate movements, and prices diverge. Investments to supply an interstate market are inhibited.

The cost, risk and facility of transition to a low-emissions economy would be improved if the transmission network of the National Electricity Market became truly national. Benefits would include:

- providing a greater geographic scope for low-emissions generator investors to select their ideal location, and many more places to which they can realistically connect
- by extending across a larger area, achieving more diversity from intermittent solar and wind resources, and also broader access for the flexible hydro generation sources in Tasmania, Victoria and New South Wales to back up the variations
- greater sharing of generation reserves, requiring less total generator capacity to meet diverse demand peaks
- enabling the market to find the most efficient source of power in a national context when a carbon price is taken into account—a high-emissions plant is less likely to be required to support local demand peaks
- providing a transmission network and market that are capable of withstanding the early retirement of carbon-intensive generation without physical or financial shocks in regional markets
- providing for more generation competition, bringing customer prices closer to an efficient level, and improving market conditions for smaller, specialist retailers.

There are also a number of detrimental effects from the lack of a truly national grid. These include inefficient overexpenditure on local transmission and distribution justified by supplying the extreme peak with reserve capacity. The National Electricity Market is made even more fragmented by the current rules, which favour intrastate regional flows over those from interstate when there is congestion across boundaries.

It is highly unlikely that a seamless national network can be built by five state-based transmission planners with parochial responsibilities. The crucial next step in transmission reform is the rationalisation of National Electricity Market transmission planning. It would not be productive to seek to merge the existing network owners, which are partially privatised. Rather, the newly created National Transmission Planner should assume responsibility for all National Electricity Market transmission planning. This would require each state to separate its transmission ownership from its planning. The Victorian experience shows that the separation is feasible. An empowered National Transmission Planner could develop single national standard charges for transmission, reliability standards and congestion management with much greater efficiency as a dividend.

It may appear to be contradictory to suggest the need for increased investment in interstate connectivity to build a truly national market, while drawing attention to excessive investment in distribution networks. There is no contradiction: rigorous assessment of costs and benefits and regulatory systems that avoid excessive rates of return on low-risk investments is required in both cases. Rigorous assessment would lead to much less investment in the near term in one case (distribution), and probably to more and different investment in the other (interstate transmission). Investment in interstate transmission would need to be paid for by some combination of budget subventions and higher prices for consumers. The cost could be minimised by drawing on investors who are comfortable with lower rates of return for low-risk investments of this kind.

Privatising distribution

There is an unfortunate confluence of incentives that has led to significant overinvestment in network infrastructure. It is clear from market behaviour that the rate of return that is allowed on network investments exceeds the cost of supplying capital to this low-risk investment. The problems are larger where the networks continue to be owned by state governments. State government owners have an incentive to overinvest because of their low cost of borrowing and tax allowance arrangements. In addition, political concerns about reliability of the network, and about the ramifications of any failures, reinforce these incentives.

A comparison of costs between Victoria, where the network providers are in private hands, and New South Wales and Queensland, where the network providers are in state hands, provides compelling evidence to support this contention (see Figure 11.3). While there are likely to be genuine differences between the states that explain some of these divergences, it is unlikely that these differences explain the majority of them.

Figure 11.3: Real capital expenditure per customer

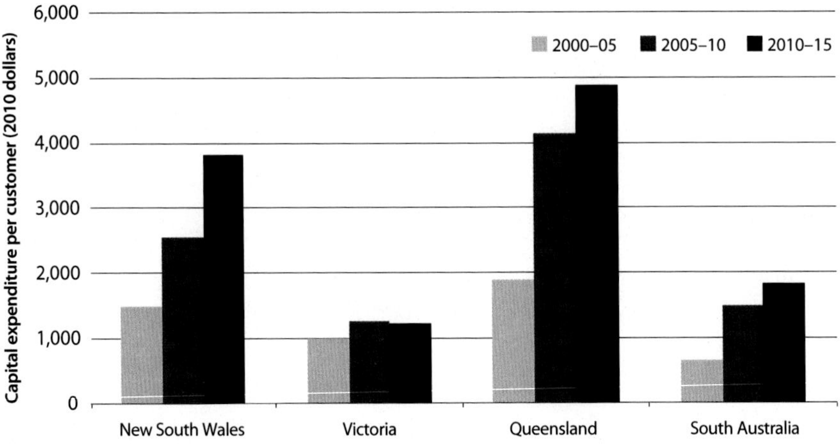

Source: Energy Users Association of Australia 2010, *Issues for the MCE: presentation to SCO meeting #59.*

Distribution networks are, of course, natural monopolies. So a strong regulatory regime is required to prevent price gouging. The Australian Energy Regulator will complete its first cycle of regulatory determinations for transmission and distribution network services providers in 2011, at which point it will take stock of the regulatory rules under which the determinations were made. This is a natural time to be considering an overhaul of the regulatory arrangements. Changes will be introduced gradually sometime after that.

But the natural cycle will lead to delays in correction of distortions that are costly to consumers and cause unnecessarily large electricity price rises at a time when the introduction of carbon pricing heightens sensitivity to them. The Ministerial Council on Energy, which is chaired by the Commonwealth minister and which supervises the regulatory arrangements, should bring forward the reform of the price regulation rules. In the states where the distortions are having the largest effects on prices—New South Wales and Queensland—state ownership of the distribution assets affects early implementation of new arrangements.

Where government ownership continues, then the rules should allow the regulator to take a different approach in regulating government-owned firms. Regulatory determinations involving government-owned firms should account for their unique borrowing and taxation arrangements.

The regulatory framework includes service standards, and providers score well against them at present. It will, however, be necessary to ensure that these schemes are appropriate to good performance in an environment in which companies are trying to reduce, rather than to increase, expenditure on the network.

The reform of price regulation must end the current encouragement of increases in peak demand for electricity, and introduce incentives for reducing peak demand.

Enter the carbon price

A carbon price will be the main driver of transformation of the electricity sector. It will alter electricity production and consumption—but it is neutral as to how this change is achieved. In some cases the carbon price may drive new investment in low-emissions generation—whether large or small scale. It may lead to fuel switching, so that established generators with high emissions run less intensively and generators that use lower-emissions fuels run more of the time. Or it may lead to the adoption of practices that lead to lower emissions from existing plants and fuels.

There will be some reduction in demand—the overseas studies suggest a 3 per cent fall in demand in the short term after a 10 per cent increase in price, and a 7 per cent increase in the long term. The easing of demand growth as prices have risen in recent years suggests the potency of these effects. Both the electricity price increases from carbon pricing and the larger increases from other sources will pull back demand.

Under a carbon price, the market, rather than the government, will be making abatement decisions, which will ensure emissions reductions are delivered at lowest cost.

With a carbon price in place, current climate change mitigation policies would not be a cost-effective way to reduce emissions. Most, including the Renewable Energy Target and feed-in tariffs, should be phased out. The Renewable Energy Target could be phased out by fixing the established price for not meeting the requirements at its current dollar level.

Modelling of the electricity sector provides an indication of a possible future mix of types of generation. There are three broad trends that can be expected. First, there is likely to be an initial increase in gas generation. Established gas plants will run for longer hours and coal plants for shorter, at relatively low carbon prices. In expectation of higher carbon prices in future, gas rather than coal will be used in new plants. The highest-cost and most emissions-intensive old plants will close at some time and the capacity that they provided will be supplied from new plants using fuels with lower emissions.

However, there is one proviso. Gas prices in eastern Australia will rise towards the price at which the same gas can be sold to overseas markets as a gas export industry develops there. In this case, the increase in gas generation may be temporarily delayed. The increase in domestic gas prices as a result of the development of an east coast export industry will be smaller to the extent that increasing global gas supplies reduce the international price.

As the carbon price rises with time, and as the costs of newer technologies fall with research, development and experience, less emissions-intensive forms of generation will become competitive. The extent of the change after movement to emissions reduction targets and a floating carbon price will depend on the cost of abatement elsewhere in the economy. With the emergence of credible international markets for abatement, the balance between domestic and international reductions in emissions will be determined by factors affecting costs of abatement in Australia and abroad.

As new low-emissions generators enter the market, supply from more emissions-intensive generators will be gradually displaced and their output gradually reduced.

The introduction of a carbon price will lower the profitability of the most emissions-intensive electricity generators. The most emissions-intensive generators in Australia are the brown coal generators located in Victoria and to a lesser extent South Australia. These generators are large contributors to baseload generation, and this role will be affected for at least some plants during the transformation to lower-emissions generation.

Industry analysts suggest that some brown coal generators are in a precarious financial position even before the introduction of a carbon price, although the profits for Australian subsidiaries reported by foreign owners suggest comfortable margins. Be that as it may, the industry estimates that, over the next five years, $9.4 billion in debt on generation assets will need to be refinanced. Approximately $6–7 billion of this debt is held by the high-emissions generators in the south-eastern states.

Part of the increase in costs from carbon pricing will be recouped by passing through the price increases to electricity users. It is not possible to say in advance what proportion of the cost increases will be passed on. This is the source of community concern about electricity price increases. But for generators as a whole, most of the carbon costs are likely to be recouped from price increases. Community concern about higher prices is the mirror image of generator concern that they will not be able to pass through costs: in the final outcome, more passing through of costs will ease adjustment pressure on generators and intensify pressure on consumers. Even with high pass-through of costs, as is likely, the introduction of a carbon price will adversely affect the financial position of the most emissions-intensive generators—those that use brown coal in generation.

When the consequences of changes in cash flows and adjustments to them are worked through in detail, it is clear that firms will have to manage financial pressures, but the risks to physical energy security are low—if not negligible. The National Electricity Market is self-correcting in terms of physical supplies; prices will rise to justify retaining capacity if the alternative is unmet demand. Furthermore, like all dynamic markets, any reduction in supply by one producer will lead to an increase in prices, which subsequently increases the profit margin for all other producers. This, in turn, provides incentives for additional investment in capacity. The most emissions-intensive plants in each region are likely to be the first to start to reduce their output and this will drive an increase in non-peak wholesale electricity prices. The owners of the next most emissions-intensive generators in a National Electricity Market region will benefit from these higher prices, and thereby be more likely to remain capable of servicing debt and generating a return to shareholders.

As in any effective market, prices and expectations of future prices will rise to keep supplies in line with demand and expectations of demand. The owners of even the most emissions-intensive physical generation assets will have an incentive to meet demand at lower output levels, possibly in an intermittent capacity, for as long as there is physical demand for the output at the prices that emerge from the market.

Assessing risks

In the debate surrounding transformation of the electricity sector, three types of risks have been commonly cited as threats to energy security.

The first risk is contract market instability. While there is an active and responsive physical spot market, transactions for electricity are primarily traded on contract markets. The many participants in the National Electricity Market have opaque contractual relationships. Retailers contract for supply in the event of high demand and thereby avoid the impact of high spot prices. With increases in the tendency for the same company to own retailing and generation, it is likely that the contract market has shrunk in recent years, but the full extent of commitments among parties is unknown.

There is some anxiety that a financial market or contract market shock or sudden unexpected change in input prices or a natural disaster or strike could lead to financial contagion, irrational behaviour and threats to energy market stability. The anxiety extends to the financial shock that could come from a participant being insufficiently prepared for the consequences of a carbon price.

If the firm operating a large and emissions-intensive generator were unable to meet financial obligations as they were due, it may be unable to reach a mutually acceptable agreement in the timeframes available. As a result, the generator would be unable to honour existing hedge contracts to retailers at a time of high spot prices. This unlikely event could trigger a financial contagion precipitating instability within the industry.

It is worth noting that such an occurrence could arise due to circumstances unrelated to the introduction of a carbon price. For example, the worst possible case of contagion risk could have been realised in the Great Crash of 2008 when the operating company of Babcock & Brown Power (now Alinta) collapsed. In that instance, the owners and lenders were able successfully to restructure their financial arrangements over an extended period of time.

There is no established mechanism within the National Electricity Market to deal with contract market instability. This is unlike regulatory arrangements in other markets, notably financial markets, in which large and negative consequences are anticipated from the failure of large firms ('too big to fail'). The electricity market is another area in which a major firm may be thought by some to be too big to fail.

The enhancement of regulatory protections in this area is warranted. This should be done through an energy security council with appropriate function. In line with understanding of best practice in the aftermath of the Great Crash of 2008, it is important that being 'too big to fail' does not protect shareholders in large enterprises from the financial consequences of changes in the business environment.

One possible safeguard against generator insolvency is a government-provided temporary energy security loan guarantee with appropriate limits. Such a device would address the transitional risk in a focused and cost-effective manner. The loan guarantee would be directed to the most emissions-intensive generators. It would be designed so as to have as close as possible to zero influence on the production decisions of owners and lenders.

A government loan guarantee on the debt of generators would have the effect of reducing the short- to medium-term probability of generator insolvency, first of all by strengthening creditor confidence. There are well-known examples of one nervous bank within a consortium causing or going close to causing a commercially sound arrangement to unravel. The loan guarantee facility would reduce the probability of such behaviour interfering with the adjustment to a carbon price.

In addition, a government loan guarantee would allow incumbent generators to refinance their generation assets at a lower rate. This would increase the chance of generators refinancing their assets on terms that maintained positive cash flows after payment of interest.

Such an energy security loan guarantee could be available to the small number of the most emissions-intensive incumbents.

The second risk suggested by some electricity stakeholders is that energy security or reliability concerns may arise from weak incentives for firms to invest in maintenance of their generators as they approach the end of their economic lives. There is potential that this could lead to sudden decommissioning of all or part of a major plant and to disruption of supply. If the only available replacement before new capacity comes online—perhaps the more intensive use of a plant designed for peaking—is higher-cost generation, this could result in sustained periods of higher wholesale prices.

These circumstances could arise with the introduction of a carbon price, although they are unlikely. To the extent that there are grounds for concern, they are more general. The same concerns could arise independently of a carbon price, as a number of large baseload generation assets approach the end of their economic lives in the coming decades and are vulnerable to an owner's financial stress. The Australian electricity market has never yet had to deal with large amounts of capacity being withdrawn from the market, with small plants being replaced by larger plants. The market has a number of ageing assets and in the future large plants would need to be replaced, thus raising issues relating to smooth transition.

There would be value in removing these concerns by augmenting the regulatory framework to deal with the increased risks of supply disruptions as plants approach the end of their lives. However, the incentives to minimise operating expenditure and delay capital expenditure on maintenance will be balanced by market incentives to continue profitable operations. Plants that cut back on maintenance levels will face higher rates of disruption, which will in turn reduce their ability to carry long-term contracts (the primary source of commercial value for all generators). Given the market incentives to undertake the appropriate level of maintenance, a light-handed regulatory approach is preferable in the first instance.

The third risk is the level of future investment in new capacity. If there happens to be a low appetite for capital investment in Australian electricity generation, the wholesale price of electricity will rise accordingly. This is a predictable and 'bankable' feature of the National Electricity Market which private investors will anticipate. In the end, the market might commence new generation a bit early or a bit late—during which prices may be temporarily depressed or inflated. This is normal for markets. In the electricity market too, imbalances between supply and demand will lead to changes in prices and expected prices, which in turn will lead to adjustments that move supply and demand back into balance. It is incumbent upon those who argue that the electricity market cannot be trusted to bring supply and demand into balance, to show analytically how the electricity market is different.

The role of households

The rapid rise in household electricity prices over the past four years has hit all households, but it tends to disproportionately affect low-income households who spend a higher proportion of their income on electricity.

While the impact of the carbon price on electricity prices will be smaller than recent and prospective increases over a number of years that have nothing to do with carbon pricing, it will still be important to understand how any increase in prices driven by the carbon price will affect households, and especially low-income households. One great benefit of market-based carbon pricing arrangements is that they generate revenue with which households can be supported without affecting the incentive to lower the consumption of emissions-intensive goods and services.

Low-income households tend to consume less energy and fuel than high-income households, but they expend a significantly higher proportion of their income on these items. The Australian Bureau of Statistics reports that low-income households spend on electricity, on average, half as much in dollar terms as high-income households, but that this is nearly double the proportion of total expenditure. These relativities have remained consistent over time.

Moreover, rural households also tend to spend proportionately more on electricity than urban households. This difference may primarily reflect the differences in average income across rural and urban households, as rural incomes tend to be lower on average.

Analysis by the Australian Treasury in 2008 predicted that in the first five years of carbon pricing, average household electricity prices would initially increase by around 20 per cent under the scenario aimed at reducing emissions by 5 per cent from 2000 levels by 2020. In this scenario, carbon prices start at about $26, rising by four percentage points per annum plus the general inflation rate. That percentage increase would have been reduced by the exceptional inflation in electricity prices since then.

There is some variation in estimates of the amount by which demand for electricity falls when prices rise. There is some evidence that the amount by which demand falls with a given percentage increase in price may be larger in some parts of Australia than is suggested by the overseas experience to which we have referred. The 2008 Review noted the significant potential for uptake of energy-efficient practices and behaviours. This potential was recently highlighted by the Prime Minister's Task Group on Energy Efficiency. But because of a number of sources of market failures, the uptake of energy-saving practices or services is suboptimal:

- information failure—the public good nature of information creates a barrier to its provision. Without sufficient information, consumers cannot make informed decisions about their purchasing choices and behaviours
- bounded rationality and capital constraints—even where people have access to sufficient information, they may make decisions that are suboptimal (for example, not paying more for a gas or solar hot water system, which will save more money later)
- split incentives, or principal–agent problems—the party who makes a decision (for example, the landlord) is not driven by the same considerations as another party who is affected by it (for example, the tenant).

While all households are likely to experience at least one of these sources of market failure to some extent, low-income households are more susceptible. For example, low-income households have relatively less capacity to pay for energy-saving products, like solar hot water or insulation, which can have significant upfront costs. Low-income households have fewer energy-consuming appliances in general, but also noticeably fewer energy-efficient appliances, and less energy-efficient homes. In low-income households insulation is less common, refrigerators are less efficient, and there is a greater reliance on energy-intensive electric heating.

A number of existing state and Commonwealth government programs address these market failures, and offer major energy and financial savings. To address information failures, such measures as energy bill benchmarking and appliance labels are highly cost-effective. Tailored energy audits, on-site implementation of simple and low-cost energy-saving measures, and ongoing advice have also achieved strong energy and financial outcomes. Regulatory standards for the minimum thermal performance and energy consumption of some fixed appliances—where benefits to the economy as a whole outweigh costs—can offer major savings.

The identification of a market failure does not in itself make a case for government intervention to correct it. One needs to be confident that the government intervention will be a cost-effective means of changing behaviour. Recent problems with Commonwealth schemes argue for caution. Any programs in future should be modest in dimension, and follow paths that have been clearly demonstrated to be successful. Some of the state government schemes seem to provide opportunities for efficient extension of support for low-income households with special electricity requirements.

Conclusion

While the challenge posed by the established reliance on coal-based electricity is large, so are Australia's opportunities for the development of alternatives at costs that are absolutely low by international standards. Australia has an abundance of high-quality resources of virtually all of the low-emissions alternative sources of energy: gas from conventional sources, coal seams and shale; wind; solar; high-grade uranium oxide for nuclear; land with low value for food that is prospective for biomass and biofuels; special opportunities for using algae in saline marine and land environments; wave and tidal energy; location adjacent to the extraordinarily rich hydro-electric potential of the island of New Guinea; geothermal from deep hot rocks; and opportunities for geosequestration of carbon dioxide. Good policy settings, including a carbon price, a fully national electricity market and reformed regulatory regimes will release these potential sources at the lowest possible cost.

In an effective global approach to mitigation, Australia would move quickly to replace high-emissions coal generation with increased output from currently operating gas plants. It would also concentrate new investment on gas and renewables, and over time would replace established coal generation capacity with new gas and renewable energy.

Transport economics will cause gas to remain the major source of energy for electricity generation in Australia for longer than in any other developed country. For the same reason, it will take longer for nuclear energy to be an economically efficient source of electricity in Australia than in any other country. Transport economics excludes nuclear from an early role in Australian electricity supply. Nuclear development in the foreseeable future would require the elevation of political preference for nuclear over the economics.

Eventually, with deeper reductions in emissions and a higher carbon price, gas itself would become uneconomic in the absence of low-cost biological or geological sequestration of emissions. Economically efficient sequestration would, of course, give coal a new economic lease on life, and prolong the economic life of gas. It seems likely that sequestration from gas combustion will be cheaper and easier than from coal.

It is not currently clear which energy sources will follow the eventual decline of gas in electricity generation. We do not need to know. It is best to keep a range of options alive. Eventual winners will depend on relative rates of technological improvement and, in the important case of nuclear, by developments in the reality and perception of its safety.

What is clear is that Australia has many attractive options for energy supply and electricity generation. If the policy settings are right, Australia will be a country of relatively low energy costs and relatively high energy use in the future, as it has been in the past. Good policy settings will provide incentives for reductions of emissions on the demand and supply sides of the electricity market, encourage innovation, and minimise the costs of transmission and distribution while fostering competition and the emergence of new supply from those low-emissions generation sources that have the lowest possible costs.

12   Choosing the future

WE CAN NOW return to the assessments of the national interest and discussion of the obstacles in the way of good policy with which we opened the book. We have to decide whether it is in our national interest to do our fair share or to lag behind others in the mitigation effort. Then we have to decide how we go about doing our fair share.

In forming our assessments of the national interest, two main questions have to be answered: Is the science legitimate? And what is the relationship between what Australia does and what the rest of the world is doing?

On the first question, the material presented in Chapter 1 confirms the central propositions from the science beyond reasonable doubt. The central propositions of the mainstream science on climate change are accepted by most Australians. This provides a basis for effective policy action.

On the second, there would be no reason to participate in a global effort to reduce greenhouse gas emissions if there were no prospect of effective global action. Chapter 4 demonstrates the reality of widespread international action to reduce emissions. This has shifted the world well below business-as-usual emissions growth. Current action holds out the possibility of evolution into strong global action that realises the international community's objective of holding temperature increases to below 2°C.

Here we are dealing with facts, not beliefs. I hope that people who oppose Australian action to reduce greenhouse gas emissions on the grounds that others are not acting will directly respond to the facts presented in this book. Differences can then be identified and additional information sought to resolve them.

Those who oppose Australian action on the grounds that others are not reducing greenhouse gas emissions can be comfortable with the recommendation in this book, that Australia strengthen its target in line with the average of international action.

There is, however, another line of argument about international action that is used by those who oppose Australian action on mitigation. This is the argument that Australia is an inconsequential country. What Australia does and does not do, according to this argument, has no effect on the actions of others. Therefore Australia should do nothing and save its money, whether or not the rest of the world is taking action. That way Australia will benefit from what others do if they are taking action, and save money if they are not.

The view that one country's actions have no effect on other countries is present in all but the largest countries, but outside Australia is recognised more clearly for what it is: an excuse for not acting on climate change. The argument dissolves once it is recognised that there is no need to make a once-for-all decision on Australia's share of an ambitious global mitigation effort. What is important is that we make it clear that we are moving with other countries, and are prepared to contribute our fair share to ambitious action if others are playing their parts. We can all build towards strong mitigation, each of us observing the actions of others and moving further in response to what we see.

What we are dealing with is a problem in which the solution requires collective action. It is not an unusual kind of problem in domestic or international affairs. Indeed, the difference between civilisation and anarchy is above all the capacity of society to find a basis for efficient collective action when it is necessary to solve a problem of great consequence.

Australians who don't want any action on climate change make the point that we account for only a very small proportion—just under 1.5 per cent of total global emissions—so that what we do has little direct effect on the global total. This is a true but trivial point. And, while the United Kingdom's share of global emissions is not much larger than ours—about 1.7 per cent, despite it having three times our population—it hasn't occurred to a British prime minister from Margaret Thatcher onwards that Britain's efforts are unimportant. And nor are they. The influence of British ideas has been considerable.

But the view that Australia doesn't matter is common enough in Australia for us to have to answer the question: is ours truly a country that doesn't count?

We could seek an answer by listening to what others say.

In Melbourne in March, the Chinese minister with responsibility for climate change policy and also energy policy, Vice Chairman of the National Development and Reform Commission Xie Zhenhua, told me that China's emissions reduction commitments would not be affected by inaction in Australia. But, he added, it was crucially important not only that Australia meet its unconditional target of reducing emissions by 5 per cent by 2020, but make the target more ambitious in line with the efforts of other developed countries. This would, he said, encourage others whose commitments were explicitly or implicitly conditional.

Xie was not saying that Australia doesn't matter.

The United States ambassador to Australia and officials in Washington reporting directly to the president have asked me not to underestimate how strongly the outcome of the current Australian policy process will feed back into the US discussion on climate change. Australia is seen as sharing some of the same characteristics of the United States, including high per capita energy use and emissions and an exceptionally large role for emissions-intensive industries in the political process. Our decision to follow the Bush administration into failing to ratify the Kyoto Protocol, after being a party to the agreement negotiations, made us the developed country whose example was cited most often in the US domestic political debate. Acceptance of carbon pricing in Australia, said the ambassador and others, would help the chances of strong mitigation action in the United States.

Count this against the doctrine of Australia as an inconsequential country.

The recognition of Australian influence is clearer and stronger in other countries in our western Pacific neighbourhood.

In the course of my work over the past four years, I have discussed climate change policy with leading members of the Indonesian cabinet on half a dozen occasions. Indonesia is certainly not an inconsequential country: it is the fourth most populous country in the world; the third largest emitter of greenhouse gases; the largest country in our region with a Moslem majority; the international policy leader of Southeast Asia; and the third biggest economic growth success story of the Platinum Age. Indonesian leaders are closely interested in what Australia might or might not do. They would be amazed to hear that some Australians think that Australia doesn't matter.

Then we can look at the historical record.

Direct experience has left me with no doubt that Australia has the standing, the analytic capacity and the diplomatic skills to significantly influence international policy on issues. When there is compatibility between the interests of Australia and the countries we are seeking to influence, and on which we ourselves are acting consistently with the shared international interest, that influence can be decisive. Climate change is such an issue.

I have been involved in a number of issues of this kind on which Australia has exercised substantial influence over the past three decades. One was persuading China that we shared a mutual interest in China developing its wool textile and steel industries with large-scale use of high-quality raw materials from abroad. A second was the influence of the idea of concerted liberalisation of trade across the countries of the western Pacific. From the middle of the 1980s until the Asian financial crisis of the late 1990s, Australia was an influential agent in the 'open regionalism' on which the Asia–Pacific Economic Cooperation forum was built during its first decade. A third was the sustained diplomatic effort that made agriculture an important focus of trade liberalisation in the Uruguay Round of trade negotiations, after neglect dating back to the beginning of the modern international trading system.

On all of these issues, Australia's influence depended among other things on our demonstrated commitment to domestic trade liberalisation—we were showing that we really believed that we shared an interest in open trade. That helped to make us influential.

There is wide recognition, in the United States and Southeast and Northeast Asia at least, that Australians are good at working out effective ways of organising international cooperation on particular issues, and at marshalling support for international cooperation around those ideas.

On the climate change issue, I would count the embodiment of 'pledge and review' in the Cancun Agreements—countries pledging their own commitments to emissions reduction and having them reviewed by other countries—as a consequence of Australian influence. 'Pledge and review' was introduced into the Copenhagen conference when it was in crisis by the Australian team, and became centrally important to President Obama's discussions with leaders of China and the other major developing countries.

And what if we applied the logic of Australia as an inconsequential country to strategic issues? Are our troops in Afghanistan, and were our soldiers on the Western Front in World War I, more influential than we could expect our contributions to shared efforts on climate change to be? Was Australia's commitment of the lives of so many of its young people in war and so much expenditure on defence over a hundred years really irrelevant to the shape of the world in which we make our lives? Would everything be exactly the same if we had decided at the beginning that our presence in Afghanistan would not affect the outcome, so that we might as well use the people and money comfortably at home?

Clearly the argument that Australia has no influence on what others do is a path into quicksand.

If the rest of the world were taking strong action to avoid dangerous climate change, and if it were true that Australian decisions were entirely inconsequential to global outcomes, would we really be comfortable to take a free ride on the efforts of others? That is not where we usually want to place our country in international affairs.

And would others be comfortable about our free riding on them, so that there was no retaliation for what others saw as inadequate contributions on climate change, and no effect on cooperation on other matters of importance to Australia?

No-one would expect the answer to be 'Yes, Australian free riding would be fine'.

Since it is not possible for Australia to be a leader in reducing greenhouse gas emissions because others are already too far ahead, we should do our fair share in what the world needs to do. Let us look forward to a future in which Australia, within the recommendations in this book, is doing its fair share in a global effort.

For some time we will be behind. That is a pity, because Australia is not an inconsequential country.

A pity, but it is where we are. And it would have a positive effect if Australia were to announce that we had established mechanisms that would allow us to catch up over time with the average effort of developed countries, and to stay there once we had caught up.

The first thing that we have to do in catching up is to make sure that we deliver the minus 5 per cent target that we have pledged formally to the international community. It would certainly damage the international effort if we looked as if we would fail. The resources boom means that we are further behind now than we were three years ago, notwithstanding the current form of the Renewable Energy Target and other policies that have been introduced over the past few years.

The introduction in 2012 of an emissions trading scheme with a fixed price on carbon for three years and then a floating price incorporating the targets that are appropriate at that time would give us good prospects of doing our fair share at moderate cost.

A starting price around $26 would make a difference to many decisions. Businesses would begin to take account of their greenhouse gas liability in a systematic way. Goods and services embodying fewer emissions would become a bit cheaper relative to other things.

We would all be aware that the carbon price would rise. This would have a larger effect on investment than on current production.

Business leaders would be on notice that we would be moving to a floating price with an emissions target in three years, so long as the conditions were there for international trade in entitlements. They would be looking at developments in other countries' carbon markets for a guide to the price at which our own system would settle as we entered into international trade. They would be calculating that it would probably be higher than the initial fixed price. They would calculate that the carbon price would be higher still if the world were becoming more ambitious in its emissions reduction targets.

Business leaders would also be aware that the price of entitlements will rise as the exchange rate falls with the end of the resources boom. Low-emissions industries are competing with the resources industries, just like our hotels, universities, factories and farms. They will become more competitive as the resources boom ends, just like the other tradeable goods and services industries. They will expand and prosper in competition with imported foreign emissions entitlements as the resources boom eases. If we have set up our institutions and policies well, the rapid expansion of our 'import replacing' low-emissions industries will be part of the increase in activity in the rest of the economy that holds up incomes and employment as the resources boom fades. Potential investors in high-emissions industries will realise that they will face greater competition as the exchange rate falls after the resources boom ends and that will discourage investment now.

If we follow the path suggested in this book, the introduction of the emissions trading scheme with an initially fixed price, and the subsequent movement to a floating price, will be important but not disruptive events in the structural evolution of the Australian economy. As soon as the parameters of the scheme are settled, business will focus on making money within the new rules, rather than on securing rules that make them money. That makes it essential that the rules really are settled. The governance arrangements proposed for the carbon pricing scheme are the key to establishing settled rules: the independent carbon bank to regulate the scheme; the independent climate committee to advise on targets and the transition to a floating price regime; and the independent agency to advise on assistance to trade-exposed industries.

Trade in entitlements would begin the convergence towards similar carbon prices across many countries. New regional trading arrangements might emerge. Each member of a regional trading arrangement of which Australia was a participant would be free to sell and buy entitlements with others so that carbon prices for our region would move closer to those elsewhere. Confidence would grow that trade-exposed industries were facing similar carbon constraints in many countries. It is likely that countries that initially chose not to take part in international trade, but which were meeting commitments to reduce emissions, would be imposing at least similar costs on their trade-exposed industries.

As the world places constraints on emissions-intensive industries—whether through the European Union's emissions trading scheme or China's direct action or in other ways—the prices of the products of these goods and services will rise more than other prices. If Australia remains a country with a comparative advantage in emissions-intensive goods when many countries including Australia are applying similar carbon constraints, we are likely to still be a large exporter of such products. The higher product prices will allow our producers to buy entitlements from abroad.

However, where there are more and less emissions-intensive ways of making the same things, the more emissions-intensive ways will lose out in the marketplace. We can be sure that if we are working within a strong global mitigation scenario, coal soon and natural gas later will cease to be competitive everywhere unless economically efficient sequestration technologies have emerged. That does not mean that we lose our advantage in energy-intensive industries. After an adjustment period in which new investment goes to various sources of low-emissions energy that are currently underutilised in the developing world, our natural advantages in a wide range of low-emissions energy sources are likely to keep us competitive as a location for energy-intensive industries.

One condition needs to be placed on the expectation that Australia will remain an internationally competitive location for energy-intensive industries in a world of strong, global carbon constraints. We have many natural advantages for low-emissions energy, but their commercial emergence will require technological innovation, skills and investment. It will require the rest of the world to see us as a good location for innovation and investment in low-emissions energy.

We will need to be seen as one of the countries that is focusing on encouraging the industries of the future and not only on protecting the industries of the past.

From the time of the floating of the carbon price, probably in 2015, the professional advice from the independent climate committee will help us to do our fair share as other countries' efforts increase. Our carbon price then rises over time as the increase in the global effort is reflected in a rising international emissions price.

We would have embarked upon the transformation towards the low-carbon economy of the future.

All Australians want to know where the new jobs and incomes will come from in a low-carbon economy. This question is related to another one: where will we find the savings in emissions that meet increasingly ambitious targets?

These were the questions that were always on people's minds as we began to reduce protection over a quarter of a century ago. The answer that my economist colleagues and I would give at the time never sounded convincing: 'From everywhere.'

'Any trade-exposed industry that is competitive in some overseas markets now', I would say, 'will become more competitive in all overseas markets, and expand. Some industries that are on the edge of being internationally competitive now will go over the edge and become exporters. Some managers in trade-exposed industries will recognise that continued competitiveness will require new ways of raising productivity, and go looking for them, and find them. Some that now depend on protection will not find ways of raising competitiveness, and their output will decline. The new jobs will be as numerous as the old jobs that have been lost.'

Well, now we know where the new jobs came from in the years after the reduction of protection. They came from everywhere. Many were in manufacturing and services industries, which increased their exports. And they were better jobs with higher pay than in the old, protected industries.

So it will be with the reductions of emissions under the market-based scheme proposed in this book. They will come from everywhere. Consumers will use less energy and other goods and services that embody high levels of emissions. Natural gas exporters will try harder to find opportunities for sequestration of fugitive emissions and the wastes from liquefaction. Landowners will think hard about the parts of their properties that would have more value as carbon sinks than they do carrying sheep. Lots of people with clever ideas of doing things in ways that reduce emissions will find equity investors and lenders more interested than they were before. Every producer will think about whether it is more profitable to spend a bit to reduce emissions, or to buy more permits.

The modelling for the 2008 Review made certain assumptions about technologies in various industries, embodying cautious views on how the technologies might change over time. That gave us one picture of how the economy would evolve if we did our fair share in a strong global climate change response, and another for our share in a weaker global effort. It showed where the emissions would be reduced to reach our targets—minus 25 per cent by 2020 and minus 90 per cent in 2050 under a strong objective—under one set of assumptions about technology. The costs of meeting our targets were manageable even if there were no extraordinary breakthroughs in technology.

What I can tell you for sure is that the outcome will not look exactly like that. It may not look much like that at all. Once we put the carbon pricing incentives in place, millions of Australians will set to work finding cheaper ways of meeting their requirements and servicing markets. We don't know in advance what the successful ideas will be, but I'm pretty sure that there will be extraordinary developments in technology. That will lower the costs of our transition to a low-carbon economy. The reductions in costs will go faster and further with the support for innovation suggested in Chapter 9.

That is the genius of the market economy. That, above all else, is why West Germany absorbed East Germany and not the other way around, and why South Korea is doing so much better than its northern neighbour. That is the reason why Australian productivity growth was so near the bottom relative to the rest of the developed world from Federation through to the mid-1980s and so high in the 1990s. That is why the United States could afford the Cold War and the Soviet Union could not. That is why economic growth accelerated in China, Indonesia and India once they had scraped away the barnacles of protection.

And that is why reliance on regulatory approaches and direct action for reducing carbon emissions is likely to be immensely more expensive than a market approach.

Direct action would have some rationale if we wanted to pretend to take action against climate change but not do much.

If we didn't do much—and remember that we would be doing quite a lot to meet our minus 5 per cent commitment to the international community—we would run great risks for Australia. It would be contrary to our national interest because it would make a strong global mitigation outcome less likely. It would be against our national interest because it would lead to our political and economic isolation and eventually to action being taken against us in international trade and other areas of international cooperation.

We would be damaged in other ways, too, if we sought to do our fair share through direct action. We would rely on the ideas of a small number of politicians and their advisers and confidants. While some of these ideas might be brilliant, in sum they would not be as creative or productive as millions of Australian minds responding to the incentives provided by carbon pricing and a competitive marketplace.

And even if the leaders upon whom direct action was relying were much smarter than the rest of us, their ideas would not be disciplined by the cold realities of the marketplace. The market has the great virtue that it quickly culls an innovation that is not proving its promise. The direct action of a politician or a bureaucrat can be kept alive on taxpayers' money, in the hope that something will turn up.

That would not be the end of the costs.

The really big cost would be the entrenchment of the old political culture that has again asserted itself after the late 20th century period of reform.

The big rewards in low-emissions investments would go to those who had persuaded the minister or the bureaucrat that their idea was worthy of inclusion in the direct action plan—if not under the government that introduced the direct action policies, then under the governments that followed. That would entrench the return of the influence of the old Australian political culture in other areas of economic policy.

And we would be leaving really difficult challenges to the Australians who follow us. In the best of circumstances we would be bequeathing them a climate that is far more difficult to live in than the one into which we were born. It would be even worse if we also left them a political culture that was incapable of the flexibility—through the use of effective markets—that will be essential to Australian adaptation to a world of climate change.

Notes

The final report of the Garnaut Climate Change Review (referred to throughout this book as 'the 2008 Review') was published by Cambridge University Press in 2008. The update to the Review, commissioned in November 2010, produced a series of eight papers on developments since 2008 and two supplementary notes, released between February and May 2011. Their titles are as follows:

Update papers (released in February and March 2011)

1: *Weighing the costs and benefits of climate change action*

2: *Progress towards effective global action on climate change*

3: *Global emissions trends*

4: *Transforming rural land use*

5: *The science of climate change*

6: *Carbon pricing and reducing Australia's emissions*

7: *Low emissions technology and the innovation challenge*

8: *Transforming the electricity sector*

Supplementary notes (released on 31 May 2011)

A 10-year plan for carbon pricing revenue

Governance arrangements for Australia's carbon pricing scheme

These materials underpin this book and provide further discussion and detail supporting the analysis. A number of commissioned reports also support the 2008 Review and the update.

The 2008 Review, the update papers and supplementary notes, the commissioned reports, and this book are all available at www.garnautreview.org.au.

Unless otherwise stated, all dollar amounts in this book are in Australian dollars.

Introduction

Page

xi   *I asked two leading econometricians*

T. Breusch and F. Vahid 2008, *Global temperature trends*, report
prepared for the 2008 Review; T. Breusch and F. Vahid 2011, *Global
temperature trends—updated with new data March 2011*, report
prepared for the Garnaut Review 2011 update.

xi   *Since 2008, advances in climate change science*

It is quite a challenge now simultaneously to respect objective truth
and to assert that there is no warming trend. A respected member of
the Australian Academy of Social Sciences rose to this challenge in
criticism of update paper 1, and a retiring senator for South Australia
and former finance minister in criticism of update paper 2. The
former's case depended on disputing the reliability of the data, and
ignored the observation of glaciers, sea-level rise and changes in
locations of plants and animals that do not depend on measurement of
temperature. The latter seems not to have felt the need to provide any
support for an argument that runs against strong intellectual authority.
See D. Aitken 2011, 'Reflections on Ross Garnaut's Cunningham
Lecture', *Dialogue* 30(1): 67–71; J. Thompson, 'Minchin ups stakes in
carbon war', *ABC Online*, www.abc.net.au, 11 March 2011.

xii   *the Great Crash of 2008*

See R. Garnaut with D. Llewellyn-Smith 2009, *The Great Crash of 2008*,
Melbourne University Publishing, Melbourne.

Chapter 1: Beyond reasonable doubt

This chapter also draws on update paper 5.

Page

1   *The vast majority of those*

A number of studies have analysed the level of agreement among
scientists that climate change is due largely to human activities, and
the credibility of scientists taking different positions on climate change.
Professor Murray Goot, Department of Politics and International
Relations, Macquarie University, conducted a review of the extent to

which the major studies demonstrated agreement among credible scientists. The review found that a range of types of evidence demonstrated that most scientists accept that human activity is a significant factor contributing to rising global temperatures. The results of the review are detailed in the following reports prepared in 2011 by Murray Goot for the Garnaut Review 2011 update: *Anthropogenic climate change: expert credibility and the scientific consensus*; *The 'scientific consensus on climate change': Doran and Zimmerman revisited*; and *Climate scientists and the consensus on climate change: the Bray and von Storch surveys, 1996–2008*.

1   *no doubt that average temperatures on earth are rising*

See, for example, Climate Commission 2011, *The critical decade: climate science, risks and responses*, Department of Climate Change and Energy Efficiency, Canberra.

2   *through the 2007 Intergovernmental Panel on Climate Change*

Created in 1988 by the World Meteorological Organization and the United Nations Environment Programme, the Intergovernmental Panel on Climate Change publishes comprehensive scientific reports about global climate change. The first review of the state of knowledge on various aspects of climate change was completed in 1990 and the latest, the Fourth Assessment Report, was released in 2007. These and other reports are prepared to inform parties to the United Nations Framework Convention on Climate Change so that climate change policy decisions are based on the best available science. Intergovernmental Panel on Climate Change 2010, *Understanding climate change: 22 years of IPCC assessment*, World Meteorological Organization, Switzerland.

3   *Carbon dioxide emissions from fossil fuel combustion*

Recent trends in carbon dioxide emissions from fossil fuel combustion are described in the following: International Energy Agency 2010, *CO_2 emissions from fuel combustion*; T.D. Keenan and H.A. Cleugh (eds) 2011, *Climate science update: a report to the 2011 Garnaut Review*, CAWCR Technical Report No. 036; M.R. Raupach and P.J. Fraser 2011, 'Climate and greenhouse gases', in H.A. Cleugh, M. Stafford Smith, M. Battaglia and P. Graham (eds), *Climate change: science and solutions for Australia* , CSIRO Publishing, Melbourne, pp. 27–46.

4   *Some recent studies have indicated*

J.G. Canadell, C. Le Quéré, M.R. Raupach, C.B. Field, E.T. Buitenhuis,
P. Ciais, T.J. Conway, N.P. Gillett, R.A. Houghton and G. Marland
2007, 'Contributions to accelerating atmospheric CO_2 growth from
economic activity, carbon intensity, and efficiency of natural sinks',
Proceedings of the National Academy of Sciences 104(47): 18866–70;
T.D. Keenan and H.A. Cleugh (eds) 2011, *Climate science update: a
report to the 2011 Garnaut Review*, CAWCR Technical Report No. 036;
C. Le Quéré, M.R. Raupach, J.G. Canadell, G. Marland, L. Bopp, P. Ciais,
T.J. Conway, S.C. Doney, R. Feely, P. Foster, P. Friedlingstein, K. Gurney,
R.A. Houghton, J.I. House, C. Huntingford, P. Levy, M.R. Lomas,
J. Majkut, N. Metzl, J. Ometto, G.P. Peters, I.C. Prentice, J.T. Randerson,
S.W. Running, J.L. Sarmiento, U. Schuster, S. Sitch, T. Takahashi,
N. Viovy, G.R. van der Werf and F.I. Woodward, 2009, 'Trends in the
sources and sinks of carbon dioxide', *Nature Geoscience* 2: 831–36;
M.R. Raupach and J.G. Canadell 2008, 'Observing a vulnerable
carbon cycle', in A.J. Dolman, R. Valentini and A. Freibauer (eds),
The continental-scale greenhouse gas balance of Europe, Springer,
New York, pp. 5–32.

4   *The magnitude and the rate of the increase*

IPCC 2007, *Climate Change 2007: The physical science basis.
Contribution of Working Group I to the Fourth Assessment Report
of the Intergovernmental Panel on Climate Change*, S. Solomon,
D. Qin, M. Manning, Z. Chen, M. Marquis, K.B. Averyt, M. Tignor and
H.L. Miller (eds), Cambridge University Press, Cambridge; T.D. Keenan
and H.A. Cleugh (eds) 2011, *Climate science update: a report to the
2011 Garnaut Review*, CAWCR Technical Report No. 036; National
Oceanic and Atmospheric Administration 2011, *Carbon dioxide
concentration trends*, US Department of Commerce.

5   *The World Meteorological Organization concluded*

World Meteorological Organization 2011, '2010 equals record for
world's warmest year', Press Release No. 906.

4   *One of the IPCC's main conclusions*

IPCC 2007, *Climate Change 2007: The physical science basis. Contribution of Working Group I to the Fourth Assessment Report of the Intergovernmental Panel on Climate Change*, S. Solomon, D. Qin, M. Manning, Z. Chen, M. Marquis, K.B. Averyt, M. Tignor and H.L. Miller (eds), Cambridge University Press, Cambridge, p. 5.

7   *Some recent work looking at events in the northern hemisphere*

P. Pall, T. Aina, D.A. Stone, P.A. Stott, T. Nozawa, A.G.J. Hilberts, D. Lohmann and M.R. Allen 2011, 'Anthropogenic greenhouse gas contribution to flood risk in England and Wales in autumn 2000', *Nature* 470(7334): 382–85.

7   *Another study used a similar approach*

S.K. Min, X. Zhang, F.W. Zwiers and G.C. Hegerl 2011, 'Human contribution to more-intense precipitation extremes', *Nature* 470(7334): 378–81.

7   *A recent study on Australian temperature and rainfall records*

A.J.E. Gallant and D.J. Karoly 2010, 'A combined climate extremes index for the Australian region', *Journal of Climate* 23(23): 6153–65.

8   *Analysis has shown that rainfall*

D. Abbs 2009, 'The impact of climate change on the climatology of tropical cyclones in the Australian region', CAWCR Technical Report.

8   *These regional findings are consistent*

M.A. Bender, T.R. Knutson, R.E. Tuleya, J.J. Sirutis, G.A. Vecchi, S.T. Garner and I.M. Held 2010, 'Modeled impact of anthropogenic warming on the frequency of intense Atlantic hurricanes', *Science* 327: 454–58; T.R. Knutson, J.L. McBride, J. Chan, K. Emanuel, G. Holland, C. Landsea, I. Held, J.P. Kossin, A.K. Srivastava and M. Sugi 2010, 'Tropical cyclones and climate change', *Nature Geoscience* 3: 157–63.

8   *a considerable body of Australian research*

B.C. Bates, P. Hope, B. Ryan, I. Smith, and S. Charles 2008, 'Key findings from the Indian Ocean Climate Initiative and their impact on policy development in Australia', *Climatic Change* 89: 339–54; W. Cai and T. Cowan 2006, 'SAM and regional rainfall in IPCC AR4 models:

can anthropogenic forcing account for southwest Western Australian rainfall reduction?', *Geophysical Research Letters* 33: L24708; W. Cai, A. Sullivan and T. Cowan 2009, 'Climate change contributes to more frequent consecutive positive Indian Ocean Dipole events,' *Geophysical Research Letters* 36: L19783; CSIRO 2010, *Climate variability and change in south-eastern Australia: a synthesis of findings from Phase 1 of the South Eastern Australian Climate Initiative*; P. Hope, B. Timbal and R. Fawcett 2010, 'Associations between rainfall variability in the southwest and southeast of Australia and their evolution through time', *International Journal of Climatology* 30(9): 1360–71.

8   *Climate models indicate that as temperatures rise*

I.M. Held and B.J. Soden 2006, 'Robust response of the hydrological cycle to global warming', *Journal of Climate* 19: 5686–99.

9   *the majority of climate models project*

T.D. Keenan and H.A. Cleugh (eds) 2011, *Climate science update: a report to the 2011 Garnaut Review*, CAWCR Technical Report No. 036.

9   *The 2008 Review noted research*

See W. Cai and T. Cowan 2006, 'SAM and regional rainfall in IPCC AR4 models: can anthropogenic forcing account for southwest Western Australian rainfall reduction?', *Geophysical Research Letters* 33: L24708; CSIRO and Australian Bureau of Meteorology 2007, *Climate change in Australia: technical report 2007*, CSIRO, Melbourne.

9   *Annual inflows to Perth's water storages*

Western Australian Water Corporation inflow data for major dams (excluding the Stirling, Wokalup and Samson Brook dams) show that annual inflow averaged 338 gigalitres between 1911 and 1974, 177 gigalitres between 1975 and 2000, 92.4 gigalitres between 2001 and 2005, and 57.7 gigalitres between 2006 and 2010, with annual inflow in 2010 dropping to 6.2 gigalitres.

9   *Analysis of historical observations confirms*

C.M. Domingues, J.A. Church, N.J. White, P.J. Gleckler, S.E. Wijffels, P.M. Barker and J.R. Dunn 2008, 'Improved estimates of upper-ocean warming and multi-decadal sea-level rise', *Nature* 453: 1090–93; M. Ishii and M. Kimoto 2009, 'Reevaluation of historical ocean heat content variations with an Xbt depth bias correction',

Journal of Oceanography 65: 287–99; S. Levitus, J.I. Antonov, T.P. Boyer, R.A. Locarnini and H.E. Garcia 2009, 'Global ocean heat content 1955–2007 in light of recently revealed instrumentation problems', *Geophysical Research Letters* 36: L07608.

9   *More recent observations indicate*

Estimates of global average sea-level rise based on observations up to 2009 are presented in J.A. Church and N.J. White 2011, 'Changes in the rate of sea-level rise from the late 19th to the early 21st century', *Surveys in Geophysics* doi: 10.1007/s10712-011-9119-1.

10   *The recent acceleration in the dynamical flow*

Developments since the 2007 IPCC Fourth Assessment Report in understanding of future sea-level change, including the contribution from icesheets on Greenland and Antarctica, are discussed in J.A. Church, J.M. Gregory, N.J. White, S. Platten and J.X. Mitrovica 2011, 'Understanding and projecting sea-level change', *Oceanography* 24(2): 84–97.

11   *a review of all observations*

For further details and a comparison of studies, see update paper 5, p. 23; S. Rahmstorf 2010, 'A new view on sea level rise', *Nature Reports Climate Change* 4: 44–45.

11   *other work suggests that a sea-level rise*

W.T. Pfeffer, J.T. Harper and S. O'Neel 2008, 'Kinematic constraints on glacier contributions to 21st-century sea-level rise', *Science* 321(5894): 1340–43.

12   *Australia's biodiversity is not distributed evenly*

W. Steffen, A. Burbidge, L. Hughes, R. Kitching, D. Lindenmayer, W. Musgrave, M. Stafford Smith and P.A. Werner 2009, *Australia's biodiversity and climate change*, CSIRO Publishing, Canberra.

13   *Measurements indicate that the average seawater acidity*

Secretariat of the Convention on Biological Diversity 2009, *Scientific synthesis of the impacts of ocean acidification on marine biodiversity*, Montreal, Technical Series No. 46.

13 *New research has focused on the tipping elements*

P. Leadley, H.M. Pereira, R. Alkemade, J.F. Fernandez-Manjarres,
V. Proenca, J.P.W. Scharlemann and M.J. Walpole 2010, *Biodiversity
scenarios: projections of 21st century change in biodiversity and
associated ecosystem services*, Technical Series No. 50, Secretariat of the
Convention on Biological Diversity, Montreal; T.M. Lenton, H. Held,
E. Kriegler, J.W. Hall, W. Lucht, S. Rahmstorf and H.J. Schellnhuber
2008, 'Tipping elements in the earth's climate system', *Proceedings of
the National Academy of Sciences* 105(6): 1786–93.

13 *In a 2009 survey of 43 experts*

E. Kriegler, J.W. Hall, H. Held, R. Dawson and H.J. Schellnhuber 2009,
'Imprecise probability assessment of tipping points in the climate system',
Proceedings of the National Academy of Sciences, 106(13): 5041–46.

14 *Simulations that incorporate*

CSIRO and Australian Bureau of Meteorology 2007, *Climate
change in Australia: technical report 2007*, CSIRO, Melbourne;
IPCC 2007, *Climate Change 2007: The physical science basis.
Contribution of Working Group I to the Fourth Assessment Report
of the Intergovernmental Panel on Climate Change*, S. Solomon,
D. Qin, M. Manning, Z. Chen, M. Marquis, K.B. Averyt, M. Tignor and
H.L. Miller (eds), Cambridge University Press, Cambridge.

14 *A recent study suggested that*

C. Tarnocai, J.G. Canadell, E.A.G. Schuur, P. Kuhry, G. Mazhitova and
S. Zimov 2009, 'Soil organic carbon pools in the northern circumpolar
permafrost region', *Global Biogeochemical Cycles* 23: GB2023.

15 *While climate change is a common driver*

A recent study ranked a number of tipping points high in both
understanding and certainty of projections (for example, the Arctic
tundra/permafrost, snow and glacier melt and tropical coral reefs).
The authors of the study concluded that while the existence of
potentially irreversible tipping points can be anticipated with
high confidence, specific thresholds cannot yet be predicted
with adequate precision and advance warning. This presents a
significant management challenge and a high risk that critical thresholds
could be breached. See P. Leadley, H.M. Pereira, R. Alkemade,

J.F. Fernandez-Manjarres, V. Proenca, J.P.W. Scharlemann, M.J. Walpole 2010, *Biodiversity scenarios: projections of 21st century change in biodiversity and associated ecosystem services*, Technical Series No. 50, Secretariat of the Convention on Biological Diversity, Montreal.

15   *Recent research suggests that solar output*

J.L. Lean and D.H. Rind 2008, 'How natural and anthropogenic influences alter global and regional surface temperatures: 1889 to 2006', *Geophysical Research Letters* 35: L18701.

Chapter 2: Carbon after the Great Crash

This chapter also draws on update papers 3 and 5.

Page

20   *what I call the Platinum Age*

The 'Platinum Age' of the early 20th century is so named because global economic growth in this period has been and is expected to continue to be more extensive and stronger than in the 'Golden Age' of the 1950s and 1960s. I used this term in December 2006 in a paper titled 'Making the international system work for the Platinum Age' for a seminar at the University of Queensland in honour of the 80th birthday of economic historian Angus Maddison. See also R. Garnaut 2011 'Making the international system work for the Platinum Age of Asian growth', in S. Armstrong and V.T. Thanh (eds), *International institutions and Asian development*, Routledge, New York, pp. 25–48; R. Garnaut and Y. Huang 2007, 'Mature Chinese growth leads the global Platinum Age', in R. Garnaut and Y. Huang (eds), *China: linking markets for growth*, Asia Pacific Press, Australian National University, Canberra.

22   *great differences in the underlying rate of change*

See, for example, D.I. Stern and F. Jotzo 2010, 'How ambitious are China and India's emissions intensity targets?', *Energy Policy* 38(11): 6776–83.

24   *The agency's most recent projections*

International Energy Agency 2007, *World energy outlook 2007*; International Energy Agency 2010, *World energy outlook 2010*.

26 *In the year to August 2010*

pitt&sherry 2010, *Carbon emissions index*, November 2010 issue.

26 *as reported to the United Nations Framework Convention on Climate Change*

The international treaty that sets general goals and rules for confronting climate change. It has the goal of preventing 'dangerous' human interference with the climate system. Signed in 1992, it entered into force in 1994, and has been ratified by all major countries of the world.

28 *One recent study analysed*

M. Meinshausen, N. Meinshausen, W. Hare, S. Raper, K. Frieler, R. Knutti, D. Frame and M. Allen 2009, 'Greenhouse-gas emission targets for limiting global warming to 2°C', *Nature* 458(7242): 1158–62.

29 *Some models have shown*

See, for example, M. den Elzen, M. Meinshausen and D. van Vuuren 2007, 'Multi-gas emission envelopes to meet greenhouse gas concentration targets: costs versus certainty of limiting temperature increase', *Global Environmental Change* 17(2): 260–80.

29 *Research suggests that the rate of uptake*

See, for example, M.H. England, A.S. Gupta and A.J. Pitman 2009, 'Constraining future greenhouse gas emissions by a cumulative target', *Proceedings of the National Academy of Sciences* 106(39): 16539–40.

29 *Some models suggest … Other models indicate*

J.A. Lowe, C. Huntingford, S.C.B. Raper, C.D. Jones, S.K. Liddicoat and L.K. Gohar 2009, 'How difficult is it to recover from dangerous levels of global warming?', *Environmental Research Letters* 4(2009): 1–9; R. Monastersky 2009, 'Climate crunch: a burden beyond bearing', *Nature* 458(2009): 1091–94; J. Nusbaumer and K. Matsumoto 2008, 'Climate and carbon cycle changes under the overshoot scenario', *Global and Planetary Change* 62(1–2): 164–72; S. Solomon, G.K. Plattner, R. Knutti and P. Friedlingstein 2009, 'Irreversible climate change due to carbon dioxide emissions', *Proceedings of the National Academy of Sciences* 106(6): 1704–09.

29 *While the timing of the climate response*

M.R. Allen, D.J. Frame, C. Huntingford, C.D. Jones, J.A. Lowe, M. Meinshausen and N. Meinshausen 2009, 'Warming caused by cumulative carbon emissions towards the trillionth tonne', *Nature* 458(7242): 1163–66.

30 *And while geoengineering has the potential*

The Convention on Biological Diversity is convening an expert group meeting in London in mid-2011 to work on defining climate-related geoengineering and assessing the potential impacts of geoengineering on biodiversity and associated ecosystem services. See Convention on Biological Diversity 2011, *Call for experts on climate-related geo-engineering as it relates to the convention on biological diversity*, notification, Montreal.

30 *A recent report looking at black carbon*

United Nations Environment Programme and World Meteorological Organization 2011, *Integrated assessment of black carbon and tropospheric ozone: summary for decision makers.*

Chapter 3: What's a fair share?

This chapter also draws on update paper 2.

Page

34 *The Australian political community*

Prime Ministerial Task Group on Emissions Trading 2007, *Report of the Task Group on Emissions Trading*, Commonwealth of Australia, Canberra.

36 *The main outcomes of Cancun were*

Pew Center on Global Climate Change 2010, *Sixteenth session of the conference of the parties to the United Nations Framework Convention on Climate Change and sixth session of the meeting of the parties to the Kyoto Protocol*, Mexico; J. Morgan 2011, *Reflections on the Cancun Agreements*, World Resources Institute, Washington DC.

41   *In addition to developments*

D. Bodansky and E. Diringer 2010, *The evolution of multilateral regimes: implications for climate change*, Pew Center on Global Climate Change; Global Subsidies Initiative, K. Lang (ed.), *Increasing the momentum of fossil-fuel subsidy reform: developments and opportunities*, IISD-UNEP Conference Report, Geneva.

42   *a 'modified contraction and convergence framework'*

The contraction and convergence approach has figured in the international climate change debate since being developed by the Global Commons Institute in the United Kingdom during the 1990s. The approach has been promoted by India and discussed favourably in Germany and the United Kingdom. Reports by Nicholas Stern and the Commission on Growth and Development in 2008 supported variations on this general approach pointing to the need for all countries to aim for equal per capita emissions over the long term.

46   *as the Australian Productivity Commission has pointed out*

Productivity Commission 2011, *Emission reduction policies and carbon prices in key economies: methodology working paper.*

46   *The Productivity Commission had to answer*

Productivity Commission 2010, *Study into emission reduction policies in key economies: Productivity Commission background paper*; Productivity Commission 2011, *Emission reduction policies and carbon prices in key economies: methodology working paper.*

Chapter 4: Pledging the future

This chapter also draws on update paper 2.

Page

48   *In April 2011 … CBD Energy*

J. Range, 'REC seller predicts a price rise', *The Australian*, 3 May 2011.

48   *To date, 89 developed and developing countries*

UN Framework Convention on Climate Change 2011, *Compilation of economy-wide emission reduction targets to be implemented by parties including in Annex I to the Convention*, Subsidiary Body for Technical

Advice and Subsidiary Body for Implementation, United Nations;
UN Framework Convention on Climate Change 2011, *Compilation
of information on nationally appropriate mitigation actions to be
implemented by parties including in Annex I to the Convention*,
Ad Hoc Working Group on Long-Term Cooperative Action under the
Convention, United Nations.

50   *The Parikh report on low carbon*

Government of India, Planning Commission 2011, *Interim report of the
expert group on low carbon strategies for inclusive growth*.

52   *Norway's emissions per person*

Excludes land use, land-use change and forestry. World Resources
Institute 2011, *Climate analysis indicator tool*, version 8.

52   *Denmark, Finland, Norway and Sweden*

Data analysis based on various World Economic Forum global
competiveness reports. See, for example, K. Schwab 2011, *The Global
competitiveness report 2010–2011*, World Economic Forum.

53   *The new targets—50 per cent of 1990 levels*

C. Huhne 2011, *Fourth carbon budget: oral ministerial statement*,
17 May. The previous target under the third carbon budget was
35 per cent.

53   *The five-year plan for 2011–2015*

China's first such plan that incorporates an emissions intensity target
in addition to an energy intensity target. Climate change mitigation
policies and outcomes for the five-year plans for 2006–2010 and
2011–2015 are discussed in W. Jiabao 2011, *Report on the work of
the government*, delivered at the Fourth Session of the Eleventh
National People's Congress on 5 March 2011; The Climate Group 2011,
Delivering low carbon growth: a guide to China's 12th five year plan.

54   *Specific fiscal interventions*

During a speech at the Australian National University in March 2011,
National Development and Reform Commission Vice Chairman Xie
Zhenhua outlined fiscal interventions including cancellation of
value-added tax rebates and application of electricity price surcharges
for enterprises with high levels of energy use. A National Development

and Reform Commission circular released in May 2010 stated that
enterprises with high electricity use in certain industries, including
aluminium, steel and cement, would be subject to surcharges of
RMB 0.1 per kilowatt hour or RMB 0.3 per kilowatt hour, depending
on levels of electricity use. These surcharges are equivalent to costs
of around $19 and $57 respectively per tonne of carbon dioxide
equivalent. Estimates are based on exchange rates current at May
2011 and a carbon intensity of energy of 0.745 tonnes carbon dioxide
per megawatt hour. Provincial governments are responsible for
implementation of the surcharges. National Development and Reform
Commission 2010, *Circular on abolishing preferential electricity
price for high energy-consuming enterprises*, NDRC No. 978 2010;
International Energy Agency 2010, *CO_2 emissions from fossil fuel
combustion 2010*.

54   *There has also been substantial fiscal support*

The Climate Group 2011, *Delivering low carbon growth: a guide to
China's 12th five year plan*; Department of Climate Change and Energy
Efficiency 2011, *Status of global mitigation action: current targets
and policies in key countries*, update of paper released by Multi-Party
Climate Change Committee in November 2010, Department of Climate
Change and Energy Efficiency, Canberra.

59   *Australia, Canada and the United States have the highest*

In relation to Annex I developed countries. Note that the following
countries have higher emissions per person: Qatar (55.5 tonnes of
carbon dioxide equivalent per person), United Arab Emirates (38.8)
and Bahrain (25.4).

60   *the Obama administration is following*

See, for example, Committee on America's Climate Choices 2011,
America's climate choices, National Academy of Sciences.

61   *A recent major study of the US gas position*

Massachusetts Institute of Technology 2010, *The future of natural gas:
an interdisciplinary MIT study*, interim report.

63   *Independent organisations have assessed*

See, for example, World Resources Institute 2010, *US climate action in 2009–10*, Washington DC. Information on carbon pricing measures in countries other than the United States and China is drawn from *An overview of international climate change policies*, produced by the Department of Climate Change and Energy Efficiency for the Multi-Party Climate Change Committee.

Chapter 5: Correcting the great failure

This chapter also draws on update paper 6.

Page

68   *As noted by Nicholas Stern*

N. Stern 2007, *The economics of climate change: The Stern Review*, Cambridge University Press, Cambridge.

71   *Modelling suggests that*

See the 2008 Review and Australian Treasury 2008, *Australia's low pollution future: the economics of climate change mitigation*, Australian Government, Canberra.

72   *The current (May 2011) price of emissions permits … The price of offsets*

CDC Climat Research 2011, *Tendances Carbone* 58: 1.

72   *The US Government recommends that economic assessments*

US Government 2010, *Technical support document: social cost of carbon for regulatory impact analysis under Executive Order 12866*, Interagency Working Group on Social Cost of Carbon, US Government.

72   *In the United Kingdom*

UK Committee on Climate Change 2008, *Building a low-carbon economy—the UK's contribution to tackling climate change: the first report of the Committee on Climate Change*, The Stationery Office, London.

72   *as Hotelling concluded back in 1931*

H. Hotelling 1931, 'The economics of exhaustible resources', *Journal of Political Economy* 39(2): 137–75.

75 *I have recommended that three independent bodies*

R. Garnaut 2011, *Governance arrangements for Australia's carbon pricing scheme*, supplementary note.

76 *As proposed by the government*

Australian Government 2008, *Carbon Pollution Reduction Scheme: Australia's low pollution future*, Commonwealth of Australia, Canberra.

Chapter 6: Better climate, better tax

This chapter also draws on update paper 6.

Page

79 *A carbon price of $26 per tonne*

R. Garnaut 2011, *A 10-year plan for carbon pricing revenue*, supplementary note.

79 *The modelling for the 2008 Review ... and the Treasury modelling*

See the 2008 Review, p. 248 and Australian Treasury 2008, *Australia's low pollution future: the economics of climate change mitigation*, Australian Government, Canberra.

79 *Other modelling has found*

S. Hatfield-Dodds 2011, *Assessing the effects of using a share of carbon price revenues for targeted tax reform: a report to the Garnaut Review 2011 Update*, CSIRO Energy Transformed Flagship, Canberra.

81 *The fringe benefits arrangements*

K. Henry 2010, *Australia's future tax system: report to the treasurer*, Commonwealth of Australia, Canberra.

81 *as in the 2011 budget*

Australian Treasury 2011, *2011–12 Budget*, Commonwealth of Australia, Canberra.

82 *This is due to a lag in the availability of data*

The Australian Bureau of Statistics has announced that it would implement a recommendation to produce more frequent (monthly rather than quarterly) estimates of the CPI if funding becomes available.

83   *Analyses in Australia, Europe and the United States*

See, for example, Australian Treasury 2008, *Australia's low pollution future: the economics of climate change mitigation*, Australian Government, Canberra; US Government 2010, *Technical support document: social cost of carbon for regulatory impact analysis under Executive Order 12866*, Interagency Working Group on Social Cost of Carbon, US Government; European Commission 2010, *Communication from the Commission to the European Parliament, the Council, the European Economic and Social Committee and the Committee of the Regions: analysis of options to move beyond 20% greenhouse gas emission reductions and assessing the risk of carbon leakage*, European Commission, Brussels; Grattan Institute 2010, *Restructuring the Australian economy to emit less carbon: main report*, Grattan Institute, Melbourne.

85   *An independent agency should be responsible*

R. Garnaut 2011, *Governance arrangements for Australia's carbon pricing scheme*, supplementary note.

86   *Table 6.1 brings together the recommended uses*

R. Garnaut 2011, *A 10-year plan for carbon pricing revenue*, supplementary note.

Chapter 7: The best of times

This chapter also draws on update papers 2 and 6.

Page

89   *Officials of the Treasury*

Commonwealth of Australia 1985, *Reform of the Australian tax system: draft white paper*, Canberra.

89   *Bob White, the president of the Business Council*

'Statement by Business Council of Australia to National Taxation Summit', *The Age*, 2 July 1985.

90   *Mr Bradley said that the Business Council*

M. Franklin, 'Julia Gillard rejects need to contain China', *The Australian*, 27 April 2011.

91 *In April 2011, Graeme Kraehe*

G. Kraehe 2011, 'Australian manufacturing—an industry sector under siege', address to the National Press Club, Canberra.

91 *Mr Howes joined Mr Kraehe*

B. Packham, 'Gillard government under growing pressure over fears of job losses under carbon tax', *The Australian*, 15 April 2011.

91 *The Governor of the Reserve Bank*

G. Stevens 2011, *Remarks at the Victoria University public conference on the resources boom: understanding national and regional implications*, Melbourne.

92 *The Australian Treasury has demonstrated*

W. Swan 2011, *Treasurer's economic note: back in the black; our patchwork economy; the carbon scare campaign; coming up*. The chairman of Bluescope Steel has commented that numbers like these do not take into account increases in the carbon price over time, or the effects on carbon pricing on the costs of other inputs, notably electricity. Systematic accounting for these influences does not seriously qualify the implications of the treasurer's note over the period to the introduction of a principled approach to assistance to the trade-exposed industries. Some members of the business community have expressed interest in moving to a principled approach to assistance administered by an independent authority like the Productivity Commission in various face-to-face meetings with me.

92 *In early May, the chairman of BHP Billiton*

This quote was part of the following statement: 'So the reality is we have to do two things at once: we have to cut carbon emissions and at the same time find ways to meet the increasing energy needs of emerging and developing economies. The sheer size of projected energy demand means that we will have to use many different sources. Each source has different costs and environmental impacts.' J. Nasser, *Address to the Melbourne Mining Club*, 9 May 2011.

93 *If baseline emissions in 2020*

Australian Government 2010, *Australia's emissions projections 2010*, Department of Climate Change and Energy Efficiency, Canberra.

95 *It still would have ended in tears*

C. Schwartz 2010, 'The Australian Government Guarantee Scheme', *Reserve Bank of Australia Bulletin*, March quarter 2010, pp. 19–26.

98 *The disciplines imposed by the Tariff Board*

See K. Anderson and R. Garnaut 1985, 'Australia's trade growth with developing economies', *The Developing Economies* 23(2): 121–37.

Chapter 8: Adapting efficiently

This chapter also draws on update papers 1, 4 and 5.

Page

100 *Let us say that the International Energy Agency*

International Energy Agency 2007, *World energy outlook 2007*; International Energy Agency 2010, *World energy outlook 2010*.

101 *A temperature increase of 4°C above pre-industrial levels*

For more information on this assessment, see the 2008 Review, p. 102.

102 *A similar argument applies to adaptation*

F. Cimato and M. Mullan 2010, *Adapting to climate change: analysing the role of government*, Defra Evidence and Analysis Series No. 1, UK Department for Environment, Food and Rural Affairs.

103 *As the Henry tax review has noted*

K. Henry 2010, 'Chapter E: Enhancing social and market outcomes', *Australia's future tax system: report to the treasurer*, Commonwealth of Australia, Canberra.

104 *But barriers to efficient water management*

R. Ben-David 2010, *Convincing regulators of the need for climate change adaptation. Really?*, presentation at the Climate Change Adaptation Workshop, Water Services Association of Australia, 25–26 October 2010.

104 *The Productivity Commission has questioned*

Productivity Commission 2011, *Australia's urban water sector*, draft report.

105 *Professor Peter Doherty asks us to consider*

P. Doherty 2009, 'Climate change/cultural change: the challenge
for the future', keynote lecture, 2009 Festival of Ideas, University of
Melbourne, 15 June.

107 *The Australian Government's assessment in 2009*

Department of Climate Change 2009, *Climate change risks to
Australia's coast: a first pass national assessment*, Australian
Government, Canberra.

107 *The recommendations from the National Climate Change Forum*

Department of Climate Change and Energy Efficiency 2010, *Developing
a national coastal adaptation agenda: a report on the National Climate
Change Forum*, Australian Government, Canberra; Coasts and Climate
Change Council 2010, *Report to Minister Combet—executive summary*,
Department of Climate Change and Energy Efficiency, Canberra.

108 *The government in 2009 identified*

Department of Climate Change 2009, *Adapting to climate change
in Australia: An Australian Government position paper*, Australian
Government, Canberra.

108 *The Council of Australian Governments' agreement*

National Emergency Management Committee 2011, *National Strategy
for Disaster Resilience: building our nation's resilience to disasters*,
available at www.coag.gov.au.

108 *This failure, combined with the vulnerability of Australian ecosystems*

W. Steffen, A. Burbidge, L. Hughes, R. Kitching, D. Lindenmayer,
W. Musgrave, M. Stafford Smith and P.A. Werner 2009, *Australia's
biodiversity and climate change: summary for policy makers*, CSIRO
Publishing, Canberra.

Chapter 9: Innovation nation

This chapter also draws on update papers 4 and 7.

Page

113 *Dr Zhengrong Shi, chief executive officer of Suntech*

Z. Shi 2011, 'Can Australia save the world?', *Focus International* 2: 14–16.

117 *the International Energy Agency has cautioned*

International Energy Agency 2011, *Global gaps in clean energy and RD&D: updates and recommendations for international collaboration*, report for the Clean Energy Ministerial, France.

118 *As leading American economist Jagdish Bhagwati has argued*

Jagdish Bhagwati has called for 'subsidising the purchase of environment-friendly technologies by the developing countries. J. Bhagwati 2006, 'A global warming fund could succeed where Kyoto failed', *Financial Times*, 16 August.

118 *Australia should commit to its share*

R. Garnaut 2011, *A 10-year plan for carbon pricing revenue*, supplementary note.

120 *The Productivity Commission opposes*

Productivity Commission 2007, *Public support for science and innovation*, Productivity Commission research report overview. See also T. Cutler 2008, *Venturous Australia: building strength in innovation*, review of the National Innovation System for the Department of Innovation, Industry Science and Research, Victoria.

121 *on-the-ground learning in Spanish solar thermal ... it is estimated that there has been a twentyfold increase*

P. Hearps and D. McConnell 2011, *Renewable energy technology cost review*, Melbourne Energy Institute.

121 *Some analysts suggest that*

See, for example, B. Balagopal, P. Paranikas and J. Rose 2010, *What's next for alternative energy?*, The Boston Consulting Group.

123 *Fugitive emissions from coal and gas*

Department of Climate Change and Energy Efficiency 2011, *Quarterly update of Australia's National Greenhouse Gas Inventory: September 2010*, Commonwealth of Australia, Canberra.

126 *There are still considerable cost-reduction opportunities*

P. Hearps and D. McConnell 2011, *Renewable energy technology cost review*, Melbourne Energy Institute.

126 *A range of sources agree*

See, for example, P. Hearps and D. McConnell 2011, *Renewable energy technology cost review*, Melbourne Energy Institute; CSIRO 2011, *Concentrating solar power—drivers and opportunities for cost-competitive electricity*, National Research Flagships, Newcastle; Geoscience Australia and ABARE 2010, *Australian energy resource assessment*, Canberra.

127 *analysis by CSIRO and others*

See CSIRO 2011, *Concentrating solar power—drivers and opportunities for cost-competitive electricity*, National Research Flagships, Newcastle.

127 *The shift to more emissions-efficient smaller cars*

S. Derby 2011, 'Green pressure: the industry is in reverse from producing unwanted vehicles', *IBISWorld Industry Report C2811: Motor vehicle manufacturing in Australia*, IBISWorld, Melbourne.

128 *Some analysts have noted that battery costs*

R. Lache, D. Galves and P. Nolan 2010, *Vehicle electrification: more rapid growth; steeper price declines for batteries*, Deutsche Bank Industry Update, 7 March.

129 *Industry projections from overseas*

P. Hearps and D. McConnell 2011, *Renewable energy technology cost review*, Melbourne Energy Institute.

Chapter 10: Transforming the land sector

This chapter also draws on update paper 4.

Page

131 *In the United States around 40 per cent of the corn crop*

The World Bank in April 2011 reported a US Department of Agriculture assessment showing that the use of corn for biofuels in the United States increased from 31 per cent of total corn output in 2008–09 to a projected 40 per cent in 2010–11. World Bank Poverty Reduction and Equity Group 2011, *Food price watch: April 2011*, World Bank.

131 *According to a recent study*

G. Fischer, E. Hizsnyik, S. Prieler, M. Shah and H. van Velthuizen 2009, *Biofuels and food security*, OPEC Fund for International Development, Vienna.

132 *A major study by the International Food Policy Research Institute*

The International Food Policy Research Institute projected food security scenarios up to 2050 using a number of different scenarios for population and income growth and climate change. The study found that food prices are likely to rise between 2010 and 2050 as a result of growing incomes and population, with additional price increases due to the negative productivity effects of climate change. G.C. Nelson, M.W. Rosegrant, A. Palazzo, I. Gray, C. Ingersoll, R. Robertson, S. Tokgoz, T. Zhu, T.B. Sulser, C. Ringler, S. Msangi and L. You 2010, *Food security, farming, and climate change to 2050: scenarios, results, policy options*, International Food Policy Research Institute, Washington DC.

136 *The agriculture, forestry and other land use sectors*

The Australian Government's most recent estimates of national greenhouse gas emissions under the accounting rules that apply for the Kyoto Protocol are provided in Department of Climate Change and Energy Efficiency 2011, *National Greenhouse Gas Inventory: accounting for the Kyoto target, December quarter 2010*, Department of Climate Change and Energy Efficiency, Canberra.

136 *The rules for the land sector*

Department of Climate Change 2008, *Carbon Pollution Reduction Scheme green paper*, Commonwealth of Australia, Canberra, pp. 119–21.

137 *The government has introduced into the parliament*

Parliament of the Commonwealth of Australia, House of
Representatives 2011, Carbon Credits (Carbon Farming Initiative)
Bill 2011, Commonwealth of Australia, Canberra.

140 *Analysis by CSIRO*

Unless otherwise indicated, references in this chapter to CSIRO analysis
relate to the following report: CSIRO 2009, *An analysis of greenhouse
gas mitigation and carbon biosequestration opportunities from rural
land use*, CSIRO, St Lucia, Queensland.

142 *Recent studies have confirmed earlier indications*

See, for example, T.D. Keenan and H. Cleugh 2011, *Climate science
update for the Garnaut Review*, CAWCR Technical Report No. 036.

143 *Recent studies have indicated substantial*

See, for example, M. Alchin, E. Tierney and C. Chilcott 2010, *Carbon
capture project final report: an evaluation of the opportunity and risks
of carbon offset based enterprises in the Kimberley–Pilbara region
of Western Australia*, Bulletin 4801, Department of Agriculture and
Food, Western Australia; CSIRO 2009, *An analysis of greenhouse gas
mitigation and carbon biosequestration opportunities from rural
land use*, CSIRO, St Lucia, Queensland; J. Sanderman, R. Farquharson
and J. Baldock 2010, *Soil carbon sequestration potential: a review
for Australian agriculture*, report prepared for the Department
of Climate Change and Energy Efficiency by CSIRO; G.B. Witt,
M.V. Noël, M.I. Bird and R.J.S. Beeton 2009, *Investigating long-term
grazing exclosures for the assessment of carbon sequestration and
biodiversity restoration potential of the mulga lands*, final report for the
Department of the Environment, Water, Heritage and the Arts by the
University of Queensland.

143 *Another assessment of abatement options*

S. Heckbert, J. Davies, G. Cook, J. McIvor, G. Bastin and A. Liedloff
2008, *Land management for emissions offsets on Indigenous lands*,
CSIRO Sustainable Ecosystems, Townsville.

143 *The West Arnhem Land Fire Abatement Project*

P.J. Whitehead, P. Purdon, P.M. Cooke, J. Russell-Smith and S. Sutton 2009 'The West Arnhem Land Fire Abatement (WALFA) project: the institutional environment and its implications', in J. Russell-Smith, P. Whitehead and G.D. Cooke 2009 (eds), *Culture, ecology and economy of fire management in north Australian savannas: rekindling the Wurrk tradition,* CSIRO, Collingwood.

144 *The area of native forest harvested*

Department of Climate Change and Energy Efficiency 2011, *National inventory report 2009*, vol. 2, Australian Government submission to the UN Framework Convention on Climate Change, Commonwealth of Australia, Canberra, p. 99.

146 *current industry estimates*

Australian Academy of Technological Sciences and Engineering 2011, *New power cost comparisons: levelised cost of electricity for a range of new power generating technologies*, ATSE, Melbourne.

146 *There is increasing recognition in Australia*

Planting and regenerating forests and woodlands can provide carbon sequestration and improve biodiversity, particularly where locally suitable native species are used. However, landholders could respond to a carbon price by favouring forests comprising a single species over biodiverse forests because lower establishment costs and higher carbon sequestration rates can make them more profitable. A study looking at southern Australia found that at a carbon price of $10 per tonne of carbon dioxide, higher carbon sequestration rates in single species plantings would lead to profits that were $7 per hectare higher than with biodiverse plantings. This example shows that incentives for biodiversity conservation that accompany the carbon price incentive could allow biosequestration activities to deliver additional biodiversity benefits at relatively low cost. See N.D. Crossman, B.A. Bryan and D.M. Summers 2011, 'Carbon payments and low-cost conservation', *Conservation Biology* 25, doi: 10.1111/j.1523-1739.2011.01649.x.

147 *Established state and federal schemes*

The Commonwealth, state and territory governments have established a range of incentive mechanisms to help protect and enhance biodiversity. Auction-based programs such as BushTender and EcoTender in Victoria have helped in expanding conservation activities on private land at relatively low cost to government. Landholders make bids based on the costs of management actions, and bids are assessed against cost and environmental benefit criteria. Landholders whose bids deliver best value for money are offered contracts and then receive periodic payments. Incentive programs can be designed to give landholders flexibility to sell biodiversity and carbon services in separate markets, and this can reduce overall costs to government. See Department of Sustainability and Environment 2008, *BushTender: Rethinking investment for native vegetation outcomes. The application of auctions for securing private land management agreements*, Department of Sustainability and Environment, East Melbourne; M. Eigenraam, L. Strappazzon, N. Lansdell, C. Beverly and G. Stoneham 2007, 'Designing frameworks to deliver unknown information to support market-based instruments', *Agricultural Economics* 37: 261–69.

Chapter 11: Electricity transformation

This chapter also draws on update paper 8.

Page

149 *In 1895, William Stanley Jevons*

W.S. Jevons 1895, *The coal question: an inquiry concerning the progress of the nation, and the probable exhaustion of our coalmines*, Macmillan, London, p. 349.

150 *After a long period in which Australian electricity prices*

Australian Bureau of Statistics 2010, *Consumer price index, Australia, Dec 2010*, cat. no. 6401.1, ABS, Canberra.

152 *Transmission network investment*

Current regulatory period revenues are forecast in regulatory determinations. Values in this paragraph are from Australian Energy Regulator 2010, *State of the energy market 2010*.

152 *Demand growth has been slow*

Data are based on the National Electricity Market, which excludes Western Australia and the Northern Territory.

154 *The Victorian experience shows*

Victoria has a private transmission network, but planning is carried out by a not-for-profit agency, the Australian Energy Market Operator. New transmission projects are competitively tendered and not subject to economic regulation.

157 *There will be some reduction in demand*

Organisation for Economic Co-operation and Development 2008, *Household behaviour and the environment: reviewing the evidence*, OECD, Paris.

158 *the industry estimates that*

Energy Supply Association of Australia 2010, *Electricity gas Australia 2010*, ESAA, Melbourne.

159 *The National Electricity Market is self-correcting*

When the system is at the point of shedding load, the price must be set at the price cap of $12,500 per megawatt hour. Or, after the equivalent of 7.5 hours of price cap in a week, an administrative price cap applies of $300 per megawatt hour.

161 *The second risk suggested by some electricity stakeholders*

Faced with large maintenance outlays and limited prospects for future revenue, owners will rationally cut back on maintenance and accept a higher risk of outage which will be traded off against the value of peak capacity. This is an intended outcome.

162 *The Australian Bureau of Statistics reports*

Australian Bureau of Statistics 2006, *Household expenditure survey 2003–04*, cat. no. 6530.0, ABS, Canberra.

162 *Analysis by the Australian Treasury in 2008*

Australian Treasury 2008, 'Chapter 6: Mitigation scenarios—Australian results', *Australia's low pollution future: the economics of climate change mitigation*, Australian Government, Canberra.

163 *This potential was recently highlighted*

Prime Minister's Task Group on Energy Efficiency 2010, *Report of the Prime Minister's Task Group on Energy Efficiency*, Canberra.

Acknowledgments

THIS BOOK is the final product of seven months of intense work by a group of able and committed people. Some worked from the start in early November 2010 to the finish in May 2011, and some for part of the time. All worked under intense pressure to meet unconscionable deadlines. On the way to this book, we produced and published eight papers and two supplementary notes, upon which this book draws. I am grateful for the sustained excellent work of wonderful people.

Steven Kennedy was head of the secretariat, sharing the development of ideas and the pressures from the beginning. His clear mind and understanding of the whole policy context added a great deal to the pleasure as well as the quality of the work.

As project director, Helen Wilson organised and held together the work of the whole team—many people, based in two cities and travelling everywhere. Over the last month she also took over the role of head of the secretariat, and so steered the project through the hectic production schedule of this final report.

I was fortunate that right at the beginning of this pressured assignment, five members of the team who worked with me on the 2008 Garnaut Climate Change Review made space within their existing commitments to help me out and provided essential continuity. Anna Freeman was director of public affairs through the seven months, building on and extending the networks of people who had expressed interest in and interacted with our earlier work. Anna organised consultations and public meetings throughout Australia and now will help me with a month's discussions with the community on the final report. Helen Morrow held together the work on the carbon pricing system. Elizabeth Edye was the anchor for the update on science. Frank Jotzo worked with me on the growth and emissions of the major developing countries. Jonathan Chew assisted on the update on innovation and electricity. Ben Skinner helped us with the electricity sector.

Paul Ryan worked closely with me on the land sector and then shared the final work on the manuscript. Anne-Marie Wilson kept us in touch with thinking on adaptation and also was part of the final production of the manuscript. Conrad Buffier helped on everything. Jane Wilkinson worked on the international issues and helped with overseas contacts.

Stuart Evans, Alice Gervink, Jenna Harris, Steven Hamilton, Thomas Abhayaratna, David de Jongh and Steven Portelli all made valuable contributions within the secretariat.

My colleagues from the 2008 Review, Ron Ben-David, Stephen Howes and Tony Wood, continued to share their wisdom through the update.

The Department of Climate Change and Energy Efficiency provided a congenial home throughout the work. Thanks to the two Secretaries during this time, Martin Parkinson and Blair Comley, and many fine officers. Thanks also to the departments of the Treasury; Resources, Energy and Tourism; Foreign Affairs and Trade; and Infrastructure and Transport; and to the Productivity Commission.

The members and independent experts on the Multi-Party Climate Change Committee discussed with me each of the update papers which were the building blocks for this final report. Among them, a special thanks to Professor Will Steffen for his advice and references to the scientific literature on many occasions.

My colleagues at the University of Melbourne provided advice and support. A special thanks to Beth Webster and her colleagues at the Melbourne Institute; the Energy Institute (Mike Sandiford, Robin Batterham and colleagues); David Karoly and colleagues in the science community; the Melbourne Sustainable Society Institute (John Wiseman and Craig Pearson); and the Monash Sustainability Institute (Dave Griggs and John Thwaites).

I am grateful for the generous support provided by scientists from CSIRO, the Bureau of Meteorology and the Centre for Australian Weather and Climate Research. Many made large contributions and I hope it does not seem exclusive if I give special thanks to Dr John Church, Dr Helen Cleugh and Dr Tom Keenan.

Thanks also to many senior members of the Australian business community for extensive discussion of the issues, including the chief executive officers of the largest mining companies and electricity generators, and in particular to Grant King in his role as chairman of the Sustainable Growth Task Force of the Business Council of Australia.

The Australian Energy Market Commission, the Australian Energy Market Operator, the Australian Energy Regulator, the Cooperative Research Centre for Greenhouse Gas Technologies, the Australian Centre for Renewable Energy and the Victorian Department of Primary Industries were valuable sources of information and advice on the energy sector.

My companion of other literary campaigns, David Llewellyn-Smith, helped me in putting together the final manuscript. Virginia Wilton and Larissa Joseph of Wilton Hanford Hanover ensured that the manuscript was clear and attractive and as free of errors as possible in a book prepared under such extraordinary pressure.

Thanks to my executive assistant, Veronica Webster, for keeping me in touch with my other professional responsibilities through my second period of deep immersion in climate change policy. And thanks to Jayne for everything.

Ross Garnaut

Melbourne

Index